D0961383

The
Evolution
of Everything

The Evolution *of* Everything

How Selection Shapes Culture, Commerce, and Nature

MARK SUMNER

PoliPointPress

14   13   12   11   10   1   2   3   4   5

Portions of this book have appeared as a series of articles
on the Web site *Daily Kos* (www.dailykos.com).

Production management: BookMatters
Book design: BookMatters
Cover design: Naylor Design

Library of Congress Cataloging-in-Publication Data

Sumner, Mark (Mark C.)
    The evolution of everything / Mark Sumner.
        p.   cm.
    Includes bibliographical references and index.
    ISBN 978-0-9824171-6-4 (alk. paper)
    1. Natural selection. 2. Evolution (Biology) 3. Darwin,
Charles, 1809–1882. I. Title.
    QH375.S86   2010
    576.8'2—dc22   2010001308

Published by:
PoliPointPress, LLC
80 Liberty Ship Way, Suite 22
Sausalito, CA 94965
(415) 339-4100
www.p3books.com

Distributed by Ingram Publisher Services
Printed in the USA

For my mother,
who tolerated dusty books,
smelly chemicals, and live lizards.
Your attention to detail and assistance is
as valuable now as it was back then.

And for my father,
who weathered many disappointments,
but was never less than proud.
Sorry to be late with this one.

CONTENTS

ACKNOWLEDGMENTS

I want to thank Markos Moulitsas Zúñiga for allowing me to use his political Web site to trot out a series of very apolitical essays and all the other editors and staff at *Daily Kos*—especially Susan Gardner, Steve Andrew, and Will Rockafellow—who were invaluable in their contributions and support.

As always, I owe a great debt to my writing group, which has (quite improbably) stayed together and supportive for over two decades of ups and downs. An appreciative tip of my hat goes to Sharon Shinn, Martha Knieb, Tom Drennan, Deborah Millitello, Laurell K. Hamilton, and Lauretta Allen.

And, finally, my gratitude goes to my wife and son, who can (at least for now) expect to get through a dinner conversation without being interrupted by my latest thoughts on how computers are like zebras.

Charles Darwin

Identity Theft

Ignorance more frequently begets confidence than does
knowledge: it is those who know little, and not those
who know much, who so positively assert that this or
that problem will never be solved by science.

CHARLES DARWIN, *The Descent of Man* (1871)

A few minutes after midnight on July 27, 1996, a security guard
working at Centennial Olympic Park in Atlanta found a green
knapsack that had been left alone near the center of the park.
Immediately suspicious, the guard contacted the police. Though
it was extremely late, the park was still swarming with locals and
with Olympic fans from around the world. The guard did not
wait for the police to arrive but at once began trying to get people
back from the suspect bit of luggage. Thirteen minutes after the
bag was first spotted, a pipe bomb hidden inside exploded. One
park visitor was killed and over a hundred more were injured. If
the guard had not acted, the toll would have been much higher,
and in reports that appeared directly after the explosion, he was
praised for his vigilance.

Three days later, the local paper began to hint that the guard, Richard Jewell, was under suspicion for the bombing. Within hours, media pundits speculated that Jewell had planted the bomb so he could pretend to discover it later and become a media hero. A former employer volunteered that Jewell was a "badge-wearing zealot." The *New York Times* referred to him as "a village Rambo." The *Atlanta Journal-Constitution* said that he "fit the profile of a lone bomber." On the *NBC Evening News*, Tom Brokaw reported that the FBI was close to making a case against Jewell. That night on the *Tonight Show*, Jay Leno compared Jewell to the serial killer known as the Unabomber, calling him the "una-doofus." A federal agent picked up the theme, calling Jewell "una-bubba." *Time Magazine* said that Jewell had "run to the limelight, and now he was frying in it."

Widespread speculation about Jewell's guilt and motivation would dominate the news for several days. At every report, charges were said to be on their way, and it was widely reported that the FBI would move against Jewell the moment it had collected enough evidence to ensure conviction.

That moment never came. The actual culprit in the Centennial Olympic Park bombing was a man named Eric Robert Rudolph, a right-wing terrorist who went on to bomb two abortion clinics and a gay night club. He viewed the Olympic Games as an expression of global socialism, and he thought that his bombing could bring the Games to a halt.

However, Rudolph was not arrested until 2003, which didn't make for neat and compelling television. Even knowledge of Rudolph's association with the crime came long after the narrative had been set. The theme of Richard Jewell as the he-caused-the-crime-so-he-could-pretend-to-be-a-hero was repeated on the airwaves so incessantly that, a decade later, far more people remember Jewell's connection with the bombing than remember the actual killer.

The bombing at the Olympics is just one example of the media shaping a story that lingers in the public conscience far longer than the truth. Whether the stories are of the Fort Meade scientist wrongly accused of the anthrax mailings, the Clintons wrecking the White House on their exit, or the "discovery" of biological weapon labs in Iraq, we've become so accustomed to the idea that the media will generate a story in the absence of facts that we almost expect it.

This phenomenon is not new. Los Angeles residents picking up the morning paper in February of 1942 could read all about the air battle that had raged over the city the previous night. The description of the battle—which included squadrons of enemy planes sweeping over the city, daring dogfights, and a fiery crash landing in a Hollywood street—had been spawned by people startled by a few shots from antiaircraft guns. Fevered imaginations, fear, and media that were anxious to "get out in front of the story" took it from there. Whether it's the Mad Gasser of Mattoon, fearful speculation over foreign terrorist cells, or the Balloon Boy, we understand that our news reflects the suspicions of our day and is often riddled with mistakes.

We don't make the same assumptions about history—but we should. History is only news at a distance, and if that means that a story has time to be reported on more frequently and more thoroughly, it also allows more time for stories to be distorted and for facts to be colored by prejudice. Worse, history is subject to two sets of distortions—those of its own time and those of *our* time—and can be inaccurate both in how it was originally recorded and in how we perceive the motivations and beliefs of those involved.

Two hundred years after his birth in 1809, Charles Robert

Darwin remains one of the most immediately recognizable men who ever lived. His image from his later years—balding and bearded, all deep-set eyes and unknowable expression—is as iconic as the *Mona Lisa*. His name conjures thoughtful discussion, heated debate, heartfelt praise, and unutterable disgust. But the ideas that made Darwin famous are almost universally misunderstood, and they have been since *before* they appeared in print.

In his bicentennial year, publishers brought out books in which Darwin was honored as the most influential man of the past millennium. Other books published in the same year derided him as a cruel racist, as a plagiarist who lifted his ideas from others, and as a marker of humanity's decline. In 2009, universities across the country held lecture series on Darwin's contributions, while other campuses played host to an effort to brand Darwin as the progenitor of Hitler and the Holocaust. Churches across the nation held special services in his honor during the week of his birth, while other churches declared his ideas "satanically inspired."

Poll after poll tells us that Darwin's ideas are either doubted or scorned by most Americans (though those polls themselves are more than a little suspect), but at the same time, Darwin remains a fixture on editorial pages and televised debates. His name has been invoked in discussions of failing banks, corporate bonuses, and global capitalism. Rejection of Darwinian ideas has become a political shibboleth for any Republican seeking public office. And, in a fight that has been waged throughout the better part of a century, school boards across the nation continue to be pressured to let Darwin into—or (more often) to squeeze him out of—America's classrooms.

For all the efforts to define Darwin at age 200, the man tends to slip through our fingers. The truth is that most Americans, including most of those who use Darwin's name for good or ill, have only the vaguest idea of either the man or his contribution to science. More importantly, they're unaware of the difference between what Darwin said and the distorted view of his ideas passed down to us by other hands.

Darwin neither invented nor discovered evolution—evolution was conceived by many others, in many places, many years before Darwin was born. Darwin didn't discredit the timeline of the Biblical account of creation; that timeline had already been cut to tatters decades before Darwin published his work. He was not the first to propose that all creatures are descended from a common ancestor, not the first to put forward a mechanism by which parents could pass their traits to offspring, and not the first to suggest that man was properly grouped among the Great Apes. All those ideas originated with others, some of them much more respected and famous in their day than the fifth child of a doctor from Shrewsbury.

So why is Darwin the one who is remembered? When not one person in a thousand has an opinion on Herbert Spencer, or Jean-Baptiste Lamarck, or Ernst Haeckel, why can Darwin's name provoke an argument as readily today as it could in 1859, when *On the Origin of Species* was first published?

Darwin is remembered for two reasons. Although others had previously floated proposals about the mechanism by which evolution operated, Darwin's hypothesis was meticulously researched, brilliantly argued, eloquently written, and vigorously defended. A hundred and fifty years of effort by the world's most motivated detractors has done nothing more than prove that Darwin's ideas were even better than he knew. Darwin is remembered because Darwin was *right*.

However, there's an even more important reason why he,

rather than some of the other historical authors of evolutionary theories, is the focus of so much attention. Darwin's ideas were dangerous. They were dangerous not just to those who counted on a rigid understanding of theology to give them purpose, but dangerous to the whole social order. It wasn't that Darwin's ideas promised to drive mankind along a path toward some dystopian ideal—it was that he threatened to topple the social pyramid. Darwin revealed that the emperor was indeed just as naked as the rest of us apes. His ideas run counter to philosophies that predate Plato and traditions older than the Christian church. His ideas were, and are, the greatest threat to the system since a Jewish healer and rabbi preached an upending of the social order in first-century Palestine.

Members of the aristocracy of Darwin's day—and of ours— were aghast at his ideas not because he left out God, but because he left out *them*. To rectify this problem, they set out to subsume Darwin, to own his name and distort his reasoning. To preserve their standing, they crafted the cult of "Social Darwinism" from a cloth far older than Darwin's writings. Like the news organizations that used Richard Jewell to create a story they could package and sell, forces uncomfortable with Darwin would carefully build a tidier, purposeful version of evolution with little regard for the facts. The cult would serve two purposes. Social Darwinism would provide a business- and status-friendly version of Darwin for the stock market, and a straw-dog Darwin ready for abuse on Sunday mornings.

These ideas would not only cloud the popular press; similar proposals would pervade academia for a century. Modern creationists seeking to dislodge Darwin from science texts like to pretend that scientists blindly followed his lead like rats after a piper, but Darwin's thoughts had their own ups and downs in the academic world. The publication of *On the Origin of Species* did not stop other theories of evolution from being presented, and some of those ideas would nearly eclipse Darwin in the decades after his death. As with the Social Darwinists, many of these theorists would borrow Darwin's name and terminology, then go on to make conjectures and conclusions very far removed from anything Darwin ever intended.

Two hundred years after Darwin's birth, and a hundred and fifty years after the publication of his most famous work, the real significance of what Darwin introduced is almost absent from public discussion. Darwin did not discover evolution, but he did reveal the extremely simple, and absolutely *inevitable,* mechanism by which evolution operates. Darwin's great idea—evolution through natural selection—is unique to biology. By showing how complexity can be generated, not from intricate plans but from the application of simple rules, he gave us a valuable gift. It's a gift that not only allows us to make sense of the story told in stones and genes, it's also a lens we can focus on economics, or culture, or ourselves.

Why was the Ford Mustang the only "pony car" to be manufactured in an unbroken string of models from the 1960s until the present? Examining possible answers through the lens of selection can provide remarkable insights. Which is a more stable economic environment: a town filled with small local shops, each of them unique and with little financial backing, or a town made up of national chain stores supported by large corporations? Traditional economics has one answer, but lessons drawn from selection may point to a very different conclusion. Why are so many consumer goods so similar? Why is bigger rarely better? Selection can provide a means of evaluating areas that appear to be unrelated to any biological process.

Because of the way Darwin's gift has been misrepresented, and because we're so often told that this gift is dangerous, we've been reluctant to take it out of the box and see what it can do. Charles Darwin is long past being offended at the slurs against his name, but *we* should be offended—offended that we've allowed this gift to be snatched away.

Those prone to discount the effects of natural selection scoff at the idea of actions that unfold over millions of years, but you don't have to wait so long to see selective pressure at work. You can see it in a day. It's in the fruit you slice onto your breakfast cereal. It's in the way you get to work. It's in your company, your children's schools, and your family church. Many of the rules that Darwin first demonstrated are with us in every action we take. We like to think that we are the masters of our own decisions, that "intelligent design" exists at least to the extent that it informs what we do, where we go, what we build, and what we believe. But there is no bright line between things shaped by human behavior and the natural world. In fact, it is *all* the natural world.

It's not my intention in this book to make a fresh argument for evolution. Evolution exists in the world as a fact, one that was recognized, discussed, and debated well before Darwin. Unwillingness to recognize evolution today in the face of overwhelming evidence stems from an insistence on a literal interpretation of creation stories (a tendency that's really quite new and that represents a radical departure from tradition) or from a deep need to see humanity as somehow separate from the rest of nature. It's extremely unlikely that I'll manage to change the mind of one person who doesn't recognize evolution, and, in fact, I'm not going to try.

I'm more concerned with revisiting the wonderful idea that Charles Darwin and the sadly neglected Alfred Russel Wallace brought to the world in 1858: natural selection. This idea generates unexpected complexity—and undeniable beauty—from the most simple of beginnings. Natural selection has been so distorted by those who have tried to make Darwin into either corporate pitchman or antireligious demon that the real implications of his work are often lost.

Rather than present a case for evolution, I want to present a case for selection itself as a force that operates not just on organisms over vast stretches of geologic time, but on everything around us from our cars and computers to our phones and food.

To examine these ideas will mean taking a trip past giant sea cows and miniature mammoths, hybrid cars and store-bought bananas, three-toed horses and sleek sea monsters. Battlefield valor and academic skullduggery. Racism, classism, more racism, great kindness, and extremes of selfishness. You may not recognize all the players at first, but they're all part of the story of evolution as it's understood by most people around the world. The story of how everyone gets it so desperately wrong.

A Portion of Hadrian's Wall

PHOTOGRAPH BY STEVEN FRUITSMAAK, 2007

CHAPTER I

Time and Money

I have this day packed a hogshead of bibles all
wrote by God's own finger.

JAMES HUTTON, correspondence (1774)

You would expect the father of economics to know a good deal when he found one. So it's not surprising that Adam Smith happily left his job as a professor at the University of Glasgow in 1763 and took up a position that had many fewer students at more than twice the pay. For three years, he served as a tutor to the young Duke of Buccleuch. In that role he traveled around Europe, met many of the leading intellectuals of the day, and helped the Duke—the future governor of the Royal Bank of Scotland—understand finance.

The job was not all that demanding, and Smith was left with time to wander through Paris and Geneva. He was allowed time to think. Time to write. In his correspondence with friends, he noted that he had begun to pull together some of his old classroom lecture notes and arrange them in a form that might eventually become a book.

When his stint as a traveling tutor ended, Smith returned to his home in Kirkcaldy, Scotland, located along the Firth of Forth, midway between Edinburgh and Dundee. By then the old lecture notes had started to take shape, and Smith had developed a theme for his text, but it would be another eight years before his notes would be published as *The Wealth of Nations*—a book that may be even more frequently misused, misquoted, and distorted than Darwin's work.

While working on his text in Kirkcaldy, Smith made frequent trips into Edinburgh to give lectures, conduct business, and meet with friends. It was there that he encountered a man named James Hutton. Hutton found Smith's economic philosophy enthralling. Trained as a doctor, Hutton had made his fortune not in the operating room but in the chemistry lab, where he had discovered a means of extracting salts used in the metalworking industry. Having been successful in two fields, Hutton turned to a third and spent a decade clearing and farming land that had belonged to his father. He had seen Smith's theories at work in all these ventures, and he was eager to discuss the topic.

Hutton was working on his own book at the time, but its topic wasn't money. It was rocks. On a rainy day in 1764, Hutton paid a visit to Hadrian's Wall. The wall had been built in the early part of the second century by a Roman emperor seeking to halt raids on settlements. South of the wall had been the areas under Roman control. North had been Celtic tribes. The wall had been the boundary of the civilized world. But what interested Hutton was not the history of the wall; it was the wall itself, in particular, the blocks of stone from which it had been made. Originally, the wall had been flanked by a ditch and topped with timber fortifications. Parts of the wall had been formed from slabs of sod. Fifteen hundred years of cold Scottish rain had reduced the ditch to a vague depression, the wood to rot, and the sod to nothing. However, the majority of the wall had been cut from blocks of pale local limestone. In places these blocks were worn, but in others they seemed as if they were fresh from the chisel. Fifteen hundred years of rain had done little to those stones.

Their remarkable preservation presented Hutton with quite a problem. He was familiar with chemical processes, and he had been making observations about the stones that surfaced in the fields of his farm. The boulders and cobbles turned up by his plow were made from different types of stones, some of which included fossils of sea creatures pressed between their fine layers. Other stones contained imprints of vanished plants. Occasionally, the stones even contained bits of other stones that had been worn down, broken, and reworked into conglomerates. Hutton visited nearby construction sites, the bluffs exposed along the banks of rivers, and the sites of nearby quarries. Everywhere he saw evidence that the stones had been weathered down, worn away, re-deposited, and formed into stone again.

But if 1500 years had done so little to the blocks in Hadrian's Wall, and materials in other ancient man-made objects were similarly unaffected, how could there have been time to so completely decompose stone that it could have been used as raw material for another generation of stones? Only a century before, Archbishop Ussher had established that Earth, in accordance with Biblical dates, was less than 6000 years old. But the stones in Hutton's fields argued otherwise. Hutton, like many other scientists around the same time, concluded through simple observation that the world was older than a strict interpretation of Biblical dates would allow. Much older.

At the time, the assumption was not just that Earth was relatively young, but that most of the features that shaped the

land were the result of a single great catastrophe—the flood of
Noah. Why were seashells found in stones located high in the
mountains? The flood of Noah. What had gouged out the great
canyons that sliced through the land? The flood of Noah. This
explanation wasn't explicitly put forward as a scientific theory.
Until the end of the Middle Ages, even the church wouldn't have
thought of interpreting the early Bible stories literally. But following the Renaissance there had been an increasing tendency
to view everything—Biblical accounts, military reports, and legal
arguments—as the same sort of data. There was no longer a sup-
position that Biblical stories were spiritual documents that spoke
of a reality different from that found on an accounting ledger.
Instead, the basic assumption was that the Bible was historically
accurate and that anything found in the stones could be made to
conform to the text. But making the physical world conform to
the scriptural words wasn't always easy to do.

Hutton wasn't the only one having such thoughts. For centuries, people had noticed different types of fossils in different
layers of stone. That observation didn't make sense if fossils had
been deposited through the action of a single great flood. Even the
stone itself varied greatly from place to place. In some areas stones
were found that appeared to have formed in water—sandstone,
shale, and limestone like that used to build the old Roman wall.
In other areas were rocks of a wholly different sort: dark and
dense rocks, like basalt; crystalline rocks, such as granite. Keen
observers saw that in some places this second kind of rock cut
across and through the layers of other stones, as if the rock itself
had been liquid at some point. In many areas, stone occurred in
layers that were quite uniform and very nearly level. Elsewhere,
rocks were sharply tilted or folded around mountains. Some layers of stone were deeply cut through by the action of streams and

rivers. Attributing all these phenomena to a single event, no matter how violent, seemed increasingly untenable.

When we look back to the eighteenth century, we're inclined to think that people accepted the Biblical origin stories from the book of Genesis in the same way that fundamentalists view these stories today, but that's not so. The sort of hard-line literal interpretation that comes with modern evangelical movements is itself a modern invention. Earlier Christians of all stripes had quite different views of what it meant for the Biblical texts to be "true," and many of those interpretations would today be viewed as radical. Thomas Jefferson famously edited the New Testament to remove miraculous events and pared the text down into more of a philosophical tract. And Jefferson was far from the most extreme.

However, even naturalists who were prepared to slash away at sacred texts had a hard time avoiding acceptance of some version of the creation story for a simple reason—there hadn't been time for anything else. So long as the first assumption was that Earth was only a few thousand years old, the second assumption *had to be* that species had appeared all at once, or nearly so, in some miraculous moment. Just as the Earth-centered idea of the universe that existed pre-Copernicus distorted every interpretation of what went on in the heavens, the young-Earth view crippled every attempt to interpret what had happened on the ground. Not just evolution but nearly every aspect of geology and biology was hamstrung until someone could deliver the thinking of naturalists from the young-Earth box. Hutton was the man for the job.

By 1776, Hutton's friend Adam Smith had published *The Wealth of Nations*, but he continued to meet with Hutton as the doctor-chemist-farmer struggled to put his ideas into simple terms. As Hutton worked, the scope of his text was continuously expanding; conciseness was not exactly Hutton's strong suit. It took 15 years before he published the first excerpts from his work, and it wasn't until 1788 that the entire work was finally published under the rather ambitious title *The Theory of the Earth*. (It would appear in several forms over the following decade, including a kind of multi-volume "director's cut" that was over 2,100 pages long and bore the even loftier title *An Investigation of the Principles of Knowledge and of the Progress of Reason, from Sense to Science and Philosophy*.)

The work was lengthy, and its reasoning was sometimes as convoluted as any mountain strata. But Hutton built his case from point to point, resulting in these conclusions:

· In areas that are currently dry land, the rocks making up the land are largely of types deposited underwater.

· Much of the material that made up those rocks was weathered away from areas that had been dry land in the past.

· What we see today as dry land is not the original land but a secondary creation made from rocks formed underwater and then uplifted from the sea.

Putting these points together revealed a cycle in which stones were raised up into dry land and then worn down by the action of wind and streams, providing the raw material for new stones that formed below the sea, which in turn were uplifted to create new land. Hutton pointed to volcanoes as a mechanism to drive the uplift of land and the creation of new mountains, and he suggested (quite correctly as it turned out) that the center of Earth was hot. This feature of his theory earned it the name *Plutonism*, after the Roman god of the underworld. The prevailing idea, that everything was created in a single great flood, was dubbed *Neptunism*.

Hutton's proposal suggested a view of the world very different from the one held by most people at the time. The drastic expansion of Earth's age created a change in perspective that was just as profound as that brought on by Copernicus, and just as discomforting. When Hutton first presented his paper, even the open-minded naturalists of Edinburgh were shocked. However, his idea confirmed the observations that naturalists—both inside and outside Great Britain—had made.

Buried among the many pages of his book was an even more radical idea. The fossils that Hutton had seen in the stones scattered around his farm, and others he had seen in a tour around Scotland, led him to speculate on phenomena beyond the formation of rocks. He theorized that life had existed throughout the cycle of deposition and uplift. During the long ages in which stones had been weathered away, deposited on the bottom of the sea, and lifted again as dry land, the world had been inhabited by creatures similar—but not identical—to those still living.

Hutton went on to propose a theory of evolution that anticipated Darwin's work on natural selection:

> In conceiving an indefinite variety among the individuals of that species, we must be assured that, on the one hand, those which depart most from the best adapted constitution will be the most liable to perish, while, on the other hand, those organised bodies, which most approach to the best constitution for the present circumstances, will be best adapted to continue in preserving themselves and multiplying the individuals of their race.[1]

This last section of Hutton's theories didn't get the public response that Darwin's work would garner decades later, partly because it was just one small part in a massive work, partly because Hutton lacked the foundation of evidence that Darwin would enjoy after another 60 years of advances in geology and biology, and partly because Hutton was not exactly a stirring writer. Immediate reviews of his work were not particularly positive, and many years passed before naturalists more generally accepted his theories.

In the end, the great gift of Hutton's work was not evolution but the time in which evolution could occur. Hutton realized that his observations of the physical world necessitated pushing the Genesis timeframe back more than 10,000 or even 100,000 years. Far more time was required to generate the effects Hutton noted. He looked on the history of Earth as an "abyss of time" that was frightening to behold, one that had "no vestige of a beginning, no prospect of an end."[2]

Without what would eventually become known as "deep time," much of modern science could not exist. Certainly evolution could not exist. Without deep time, the only option was to believe that each creature had been the subject of special creation. It might not be a scientific theory, but without deep time supernatural intervention was the *only* alternative. A scientific theory of evolution could not exist without the time Hutton provided.

Eventually, many of Hutton's theories would be validated, and he would come to be regarded as the father of modern geology. The contributions of Hutton's friend Adam Smith were recognized more quickly. While Hutton was struggling toward publication, Smith's book—which had the advantage of being much more accessible to the general public than Hutton's two-volume work—became a best seller, and the economist's acclaim was widespread. Smith was made a fellow of the Royal Society of London, a founding member of the Royal Society of Edinburgh, and Lord Rector of the University of Glasgow. The ideas in Smith's book were readily embraced by philosophers and governments the world over.

This relative difference in their success didn't shake the friendship of the two men. When Adam Smith died in 1790 after a brief but painful illness, he named James Hutton as one of the executors of his unpublished works. So the father of modern geology went through the manuscripts of the father of modern economics, determining which of the remaining notes needed to be brought to the attention of the public.

For two decades in Edinburgh, natural science and economic science talked together, dined together, and were forged together in friendship. Unfortunately, for both science and economics, the relationship would not always be so congenial.

The Great Chain of Being

RHETORICA CHRISTIANA, 1579

CHAPTER 2

And All's Right
with the World

> If some race of "four-handed" animals, especially one of the
> most perfect of them, were to lose, by force of circumstances
> or some other cause, the habit of climbing trees and grasp-
> ing the branches with its feet . . . these four-handed animals
> would at length be transformed into two-handed animals,
> and the thumbs on their feet would cease to be separated
> from the other digits, when they only used their feet for
> walking.
>
> JEAN-BAPTISTE LAMARCK,
> *Philosophie Zoologique* (1809)

If you happened to have a headache, and if you happened to live
in the city of Ur around 4500 years ago, there's a good chance you
would get a treatment that's not altogether unfamiliar. Records
from ancient Sumer include the description of a tea made from
willow bark that was given for both aches and fever. In the fifth
century BC, the Greek physician Hippocrates dispensed the
same prescription. Several tribes of Native Americans used a

similar remedy. Around the world and up until the nineteenth century, physicians offered this infusion as a nostrum against disease and disorders.

In the first decades of the 1800s, several scientists extracted salicylic acid from willow bark and determined that this was the compound that made the tea so effective. In 1853, a French chemist first prepared the related compound acetylsalicylic acid. Forty-six years later, Bayer Drugs and Dye dubbed the substance "Aspirin" and began to sell it as a remedy for a number of ailments.

From the start, there was little doubt about Aspirin's efficacy against a host of problems. It was considered a genuine wonder drug during the worldwide flu pandemic of 1918, when those who moderated their fever with the white tablet were more likely to survive. It was considered so important that the patent on the drug was forced from Germany as part of the war reparations at the end of World War I, when it stopped being "Aspirin" and became simply "aspirin." There was only one thing seriously wrong with aspirin: no one knew how it worked.

It would be another six decades before a British pharmacologist, John Robert Vane, was able to show that aspirin suppresses the production of two groups of fatty acids, prostaglandins and thromboxanes. Prostaglandins help shuttle pain messages to the brain. Thromboxanes help regulate the clotting of blood. By reducing the levels of both, aspirin battles headaches, inflammation—and heart attacks.

John Vane was not the man who discovered aspirin. He was the man who discovered how it worked.

Vane won the Nobel Prize in 1962 for his work on ferreting out the way this most familiar drug imposes its best-known actions. However, Vane's work wasn't the last word on aspirin. Since then hundreds of researchers have explored the complex interaction of this drug and living systems. They've refined knowledge about the paths by which fever and pain are controlled and looked into how aspirin reduces the production of radicals in the body, how it changes the ratios of hormones in the pituitary, how it alters the activity of the mitochondria within every cell of the body.

When it comes to evolution, Charles Darwin is not the equivalent of the ancient Sumerians or Hippocrates or Bayer. He's not the man who first discovered evolution. He's not the man who put the idea of evolution into general circulation. Darwin is John Vane. He's the man who showed how evolution works. Like Vane's work, Darwin's has been extended and refined, and like Vane's subject, Darwin's had a long history. Like Vane, Darwin has been proven correct by further research.

But there's one big difference—no one had a huge emotional investment in the mechanism behind aspirin. When Darwin published *On the Origin of Species by Means of Natural Selection*, his detractors were not just those who opposed the whole idea of evolution. Some of his most ardent attackers were evolution's most vocal supporters. Darwin's work was intellectually compelling and his arguments difficult to refute, but it was all very, very unsatisfying. And there were alternatives.

The foundations for those alternatives were as old as willow bark tea. Creation stories around the world include the organization of creatures into categories. Genesis is not unique in establishing groups of plants and animals, birds and fish, mammals and "creeping things." Quite likely, this inclination to slot creatures into categories is not solely human but is itself something inherited from our predecessors—and something we share with many other species still in the world today. When one is able to say, "that snake is very like that other snake, the one with such a painful bite" or "this plant is similar to the one we ate last week, which was good," the value is so great that it goes without notice. The naming of species—the first task God assigns to Adam—is vital, obvious. Innate. And classifying plants and animals is a prerequisite for any theory of evolution.

We recognize that animals exist as individuals, as members of a particular type (for example, "quail"), and as examples of a broader type ("birds"). With that general understanding, it's not too surprising that people of many different traditions have had ideas about the relationship of creatures and have pondered the question of whether the various types of animals were a fixed set or they changed over time. A hundred years before Hippocrates gave his prescription for willow bark, the Greek philosopher Anaximander speculated that all animals descended from a common ancestor. Similar thoughts were expressed by philosophers from Arabia to China.

In medieval Europe, the church dominated most thinking, and the thinking of the church was a mixture of Hebrew texts, Christian scriptures, and the philosophical teachings of the Greeks. The most commonly expressed relationship between plants and animals took on a form that was a hybrid between scriptural interpretation and the teachings of Aristotle. The "Ladder of Nature" or "Great Chain of Being" mapped out tiers of existence from the insensate state of rocks and minerals through the plants, animals, and men—and on to angels and God. Within each of these broad levels, there were finer divisions—layers within layers—where individual species were arrayed.

The critical feature of this arrangement is that creation isn't just grouped, it's *ranked*. Each tier is intrinsically better than the tier below. Animals are not just like each other more than they are like plants, they are *better* than plants because of their increased ability to move and experience the world. Even within a tier, the arrangement is intended to denote relative value. Lions and sheep are both mammals, but lions are better than sheep. Pigeons and hawks are both birds, but hawks are better than pigeons. Not only is man one tier closer to the angels and God than the other animals, but men themselves are ranked by their culture, race, and place in society.

It's not surprising that this idea would resonate with people at that time. Both in the church and in the society, every person was locked into a position that was almost as fixed as the stars. The Great Chain is an expression of feudal society writ large; every aspect of creation fit neatly and fixedly into a chart of God's kingdom. This hierarchical view of society felt right and hardly feels less so today. Think of any other type of grouping you can imagine—companies, sports teams, cars, occupations—is it possible to construct such a set without involving some idea of which one is "better?"

Not only people of past centuries find this type of structure familiar. If the Chain of Being takes feudalism and extends its structures across all creation, the modern corporation (as Douglas Rushkoff noted in his book *Life, Inc*) is little more than a kind of feudalism-to-go. We may substitute the term *worker* for *peasant,* use *vice-president* in place of *baron,* and spend our days at the whim of the CEO rather than the king, but it's still a system designed to extract labor from the bottom of the chain and funnel rewards to those at the top. "Moving up" or reaching a "higher rank" is as central to our lives today as it was to any aspiring nobleman of the past.

The first theory of evolution that gained widespread support during Darwin's time also had a sense of direction. The theory that so excited scientists at the start of the nineteenth century wasn't Charles Darwin's theory of natural selection. Darwin's ideas would not be published for decades. Instead, the greatest enthusiasm was directed toward the ideas of a retired French military officer and scientist, Jean-Baptiste Pierre Antoine de Monet, Chevalier de la Marck, more commonly known as Lamarck.

Born into a family with more titles than money, Lamarck had left university, ridden hard to the battle front, and joined the French Army in the Pomeranian War. The unit to which Lamarck was assigned almost immediately came under a heavy artillery barrage that killed most of the men in the unit—including all the officers. The other volunteers begged Lamarck, the only aristocrat among their ranks, to accept command of the unit and order their retreat. The 17-year-old Lamarck took command as they asked, but instead of retreating he ordered that they hold their position no matter what. They held. The tattered remains of the unit survived the day, and afterward Lamarck was given a battlefield promotion to officer's rank.

Lamarck was uninjured in the fighting, but the celebration nearly killed him. In the aftermath of his promotion, one of his comrades gave the new officer congratulations so hearty that it landed Lamarck in the hospital for weeks. (You have to wonder if the over-enthusiastic hug came from one of the fellows Lamarck ordered to stay on the field.) Afterward Lamarck was given a desk job. His battlefield heroics began and ended on the same day.

On leaving the army, Lamarck studied medicine, botany, and anything else that captured his energetic mind. The enthusiastic young man soon came to the attention of the French Academy of Sciences, and within a few years he was traveling the world, collecting plants for the royal gardens. Year after year, Lamarck rose in rank and prestige. By the time the French Revolution came along, he was in charge of a portion of the royal gardens—which he deftly renamed when association with the word "royal" was no longer such a healthy idea. Lamarck not only survived the revolution, but thrived, becoming a professor of natural history and a curator of the new natural history museum.

By 1800, Lamarck was 55 years old. He was well known, an acknowledged expert in botany, zoology, and all aspects of natural history. He was charming. He had friends in high places. He enjoyed a level of respect close to that of his countryman and rival, Georges Cuvier. But for all his vaunted knowledge, Lamarck had still published almost nothing. That was about to change.

In a three-year period, Lamarck overthrew the existing classification of animals. He was the first to pay enough attention to all those "creeping things" to correctly split off spiders and their kin from the six-legged insects and the first to establish many of the groups we know today among the invertebrates. He created a system of geology that recognized Hutton's great expanse of time and included a form of continental drift (which Lamarck thought was caused by the effects of erosion and deposition from ocean currents). He introduced the term *biology* for the general study of living systems. And he outlined his ideas for the evolution of all life from common ancestors.

As it became obvious that extinction was widespread and that whole orders of beings had vanished over time, the need grew for some theory that would explain how new species were created. Many, many things were gone. And yet the world still held abundant and diverse forms of life. Cuvier and others had shown that extinction was not only possible, but common. The rocks were full of creatures that had passing resemblance to the animals still living, but they were distinctly different animals. There were no mastodons or woolly rhinoceros roaming eighteenth-century France. But . . . France was not empty. If the grass wasn't being cropped by elephant teeth, it was still being cropped by something. The seas that had once abounded in the odd pill bug–like trilobites boasted no more of their kind, but the seas were still swarming with life.

Clearly Earth didn't contain the same mix of creatures it had held in the past, but the roles those extinct creatures had once played had been passed to new organisms. If many things had vanished . . . where had their replacements come from? Without a doubt animals had adapted. Populations had shifted. Evolution *had* occurred. But how?

Lamarck wasn't the only one writing about the issue (Charles Darwin's grandfather was one of those who had speculated on evolution some years earlier), but Lamarck was the first to provide a systematic answer. Observing that the oldest rocks contained only simple organisms, Lamarck inferred that there was a force that compelled life to grow more complex over time. As a mechanism for this force, Lamarck proposed that high-pressure fluids circulating through the body (blood, lymph, and so on) were constantly forging new paths, and that this led to the formation of new organs. Organs and tissues that got a lot of use also got more fluid, so these areas were most likely to be affected by an increase in complexity.

The part about the fluids sounds a little silly now and probably did then too. Certainly it's not cited by many of Lamarck's admirers. But other aspects of the theory made it possible to overlook these details.

Seeing the way that animals were well suited to their environment, Lamarck postulated that another force split species to make them more suitable for differing circumstances. Lamarck viewed this force in a way that any gym rat would recognize as an extended version of "use it or lose it":

> If an animal hasn't reached the limits of development, frequent use of any organ will strengthen, develop and enlarge that organ, with development matching the amount of use. Not using any organ gradually weakens it, diminishes its function, and finally causes it to disappear.[1]

Spend your time pumping iron, you get muscles. Spend that same time at a desk, and you get (or at least, I get) more gut than biceps.

Lamarck theorized that these adaptations, acquired through a life of using body parts more or less, were passed along to children. So your efforts in the gym—or the library—would give your children a head start in the same area. This part of his theory his contemporaries liked. This part they liked *very* much.

The evolutionary theory that Lamarck introduced in 1800 (and expounded on over the next two decades) provided everything you could possibly want in a theory—especially if you were intelligent, healthy, wealthy, and well placed in society. Here was a proposal that said there was a drive toward more complex, "improved" organisms and that achievements of all sorts were passed along to children.

So if you happened to be reading his theories in the large, warm library of a well-appointed home, wasn't it because your ancestors had bequeathed you not only their bank accounts but the physical and mental advantages they had labored hard to attain? Wasn't any elevation above the mean deserved? And didn't this prove that those of the lower classes—the unlovely, unhealthy, uneducated, and most certainly unwealthy masses— were in that state precisely because their ancestors had failed to put in the accumulated work to lift their lot? Class wasn't just a position, it was an accomplishment.

Lamarck himself was never too keen on the broad application of his theory to social classes, but that didn't halt its popularity. By the middle of the nineteenth century, this you-have-what-you-deserve view of evolution was being applied not only to biology, but in all fields—including economics and sociology.

It shouldn't be too surprising that when the teenage Darwin was sent to the medical school at the University of Edinburgh—as much as a companion for his older brother as for his own education—he ran into several professors who were fans of Lamarck and his theory. One of his professors wrote a lengthy (though anonymous) paean to Lamarck that was published in a local journal. Darwin recorded at least one instance in which one of his professors waxed ecstatic over the beauty of Lamarck's theory and its ability to explain so much of the variety found in the natural world. Darwin, however, did not seem to be particularly impressed. Mostly, he seemed to alternate between being bored with medical lectures and being disgusted with the practice of medicine. Watching actual surgery, which at the time was performed without anesthesia and required both speedy use of a saw and an operating theater with a floor well coated in sawdust to absorb the blood, left Darwin feeling ill. Listening to lectures on geology, which covered Hutton's idea of deep time and the struggle between Plutonism and Neptunism, interested him only slightly more than the repetitive tasks of anatomy.

In truth, Darwin acted like what he was—a sixteen-year-old being forced to study for a career that was more his father's idea than his own. While his older brother Erasmus diligently attended to his lessons, Charles Darwin enjoyed his time out of doors collecting fossils, shells, and marine life from the Edinburgh shore. His class work he simply neglected. By the end of 1827, Darwin's father grew so frustrated with his youngest son's poor study habits that he pulled Charles from medical school and sent him off to Christ's College, where he could study to become a parson. Darwin seemed not to mind. He could wander around out of doors in Cambridge just as well as in Edinburgh, and there were whole new sets of bugs and plants to be gathered.

At Cambridge, Darwin finished his bachelor's degree. Either he found the teachers there more inspirational or he found himself more ready to concentrate. The geology that had so bored him at sixteen seemed more interesting at twenty. Many of his professors had studied the ideas of Lamarck, but there were other ideas as well. Mostly what Darwin seemed to learn at Cambridge was a desire to learn more. He could see that there was a revolution under way that ran through biology, geology . . . in fact, through all of science. He wanted to contribute. He wanted to see the world. And he soon got the chance.

Meanwhile, Lamarck's ideas were already shaping the world. To a larger extent than most people realize, they still are.

HMS Beagle on a Phosphorescent Sea

CHARLES DARWIN, HARPER, 1879

CHAPTER 3

Uniformly Catastrophic

In reply, I can only plead that a discovery which seems to contradict the general tenor of previous investigations is naturally received with much hesitation.

CHARLES LYELL, *The Antiquity of Man* (1863)

In the first week of October 1836, Captain Robert FitzRoy married Mary Henrietta O'Brien, the daughter of a prominent army general. Within a few years, the respected naval officer would become a member of Parliament, then the governor of the British colony of New Zealand—a post he would lose by being so impolitic as to treat the Maori on a basis of equality. In his public career, FitzRoy would become the head of the British Meteorological Department and be heralded as the father of modern weather forecasting. He would invent a better, cheaper form of barometer and devise the first system of storm warnings. He would be the commander of one of the first steam vessels in the British navy and the master of the dockyard at which the most innovative ships of the day were constructed. He would invent new techniques of cartography, explore distant lands, and lead some of the most celebrated expeditions in history.

And his fame would be completely eclipsed by that of a man he took as a passenger on one of his voyages—a man who had recently finished off a bachelor's degree and was recommended by a friend to FitzRoy as a good conversational companion—a young fellow named Charles Darwin.

Even the book we know today as *The Voyage of the Beagle* was originally the third volume of a set commissioned to record the expeditions led by FitzRoy. Captain FitzRoy himself wrote the first two volumes. Darwin, again at FitzRoy's invitation, wrote the third. It's that third volume that's still in print more than a century and a half later.

Both men were adventurous, intelligent, innovative, and well connected. Both were very young. (FitzRoy was only 26 when he wrote seeking someone with more knowledge of the natural world to come along on his next journey. Darwin was 22.) Over the course of a journey that was supposed to take three years but ended up lasting for nearly five, the two men talked, argued, and used each other as sounding boards. They suffered through periods of cold silence and enjoyed occasions of warm friendship. FitzRoy—not only a ship's captain but the nephew of a duke—could be high-handed, officious, and prone to fits of rage. His crew nicknamed him "Hot Coffee." More than once he banished Darwin from his sight (as when the two argued over slavery, which Darwin strongly opposed) only to apologize sincerely a few hours later.

Darwin escaped to the land for over three years of their journey, leaving FitzRoy to carry out the expedition's real purpose: creating detailed maps of South America's coasts and harbors to aid British shipping. But on many occasions FitzRoy delayed the progress of the ship to allow Darwin his observations and excavations, and when possible the two men went off together to share adventures. They were friends.

The two men mused over concepts in theology and worked through scientific theories to explain the features of the world they observed together. It was during one of these discussions that FitzRoy and Darwin came to an agreement: the world around them could not have been formed in the short time allowed by a traditional interpretation of Biblical events. In particular, the placement of fossils and the structure of the land could not have been the result of the single great flood relayed in the story of Noah.

But that moment of agreement came before the wedding.

All during their long journey, FitzRoy held one personal secret that he never shared with Darwin—he was engaged. Within days of the *Beagle's* return to England, FitzRoy married the general's daughter in a wedding that was a complete surprise to Darwin. The wedding drove a wedge between the two men. As it happened, FitzRoy's new bride was a devout churchgoer, and lest he offend her, the captain censored the "progressive" ideas he had discussed during the journey. So, despite what he had told Darwin, FitzRoy's volumes recording the voyages of the *Beagle* were careful to present the results of landforms they had seen in terms of how they might have been gouged out by the waters of a single universal flood. Even seashells found on the highest peaks were explained in terms of Noah's deluge.

This school of geologic thinking came to be known as *catastrophism*—the idea that the world had been primarily shaped by a massive, unique event. Though catastrophism originated among those who traced natural history through Biblical stories of creation, it wasn't restricted only to those who expected the world to reflect those beliefs. In fact, the best-known proponent of catastrophism at the time was also the best-known scientist in the world, period—French anatomist, paleontologist, and biologist, Georges Cuvier. Unlike FitzRoy or his rival Lamarck, Cuvier had no ancestral wealth or position in the peerage. A polymath genius whose ideas were shaped by the Enlightenment, the Frenchman didn't come to catastrophism as a means of explaining what he had read in the Bible. He was looking for an answer to what he found in the rocks. In particular, Cuvier was looking for an explanation for something that was only then becoming widely accepted—extinction.

Meanwhile, the man who had just finished collecting some of the best evidence in support of extinction's worldwide reach had a completely different idea of how Earth had been made. The biggest influence on Charles Darwin's geologic thinking was another unlikely character—a former lawyer who had taken to studying rocks when his failing eyesight made it too difficult for him to read the intricate text of legal documents. Charles Lyell's *Principles of Geology* would not only shape much of Darwin's thoughts but dominate geology for the next . . . ever. In this three-volume set, Lyell laid out rules that seem simple (and are) but are also as essential to geology as a number system is to mathematics.

What Lyell taught can be boiled down into a single, intensely profound statement: the present is the key to the past.

Want to understand how ripples were formed in the sands of ancient stones? Look for the places where the same kinds of ripples are being formed on sands today. The world of the past wasn't shaped by extraordinary events, but by extremely ordinary events. Want to see what made the world? Look at wind. Look at rain. Look at streams. Look at the sea. The only difference between what happened then and what's happening now is time, lots and lots and lots of time in which those streams and oceans had the chance to carry out their actions. Our world was formed by the slow, steady accumulation of tiny modifications that led to large changes only over the course of millennia. This was a theory called *uniformitarianism.*

Lyell's volumes drew attention to the feature that had first appeared in Hutton's work—the huge amount of time required to account for geological features. But Lyell's book was more popular, more accessible to the general public, and presented much greater evidence of Earth's vast age than had Hutton's. As a result, it drew much more ire.

In the decades after the Enlightenment, science was becoming increasingly more professional, but there were still a good number of part-time naturalists, including a group of clergymen who dabbled in geology. These "Scriptural Geologists" made every attempt to fit their discoveries into their interpretation of Biblical accounts, and they attacked Lyell for his "atheism." But Lyell considered himself a devout Christian. Science and religion, said Lyell, were different kinds of truth. If a scientist grew excited about evidence that seemed to fit with scripture but ignored evidence that did not, then he was being a very poor scientist and a poor Christian. A good scientist should interpret the evidence before him as if the scriptures did not exist. And a good Christian should not find his faith threatened by science. This position has been held by many scientists from Lyell's day to the present, and it was gradually adopted by most Christian churches.

The attraction of uniformitarianism was great, and completely understandable. Uniformitarianism turned natural history into a testable, predictable endeavor. In short, it makes geology into a science. A world built completely around catastrophism would be a world of nightmarish, unpredictable events whose record can only be puzzled over—like a film made up of random clips. Uniformitarianism is a guide by which past events can be interpreted sensibly. It became the foundation for geology and paleontology, and the basis of Darwin's theories.

The idea that stones and bones are both shaped by step-by-step change became such a primary feature of Darwin's thinking that he defended uniformitarianism against those who proposed that more abrupt changes gave rise to the differences among species.

> Mr. Mivart is further inclined to believe, and some naturalists agree with him, that new species manifest themselves "with suddenness and by modifications appearing at once." . . . He thinks it difficult to believe that the wing of a bird "was developed in any other way than by a comparatively sudden modification of a marked and important kind," and apparently he would extend the same view to the wings of bats and pterodactyls. This conclusion, which implies great breaks or discontinuity in the series, appears to me improbable in the highest degree.[1]

When Cuvier died in 1832 (in the midst of Darwin's prolonged trip on the *Beagle*), catastrophism ruled the day. For years after, respect for Cuvier's genius and the support of the church helped to keep the theory at the forefront of scientific thought. In fact, the cult of Cuvier strengthened after his death when it was discovered that his brain was found to be extraordinarily large. This bit of trivia was often cited by those who wanted to make a direct connection between brain size and intelligence, as well as those who misused Darwin's theories to promote ideas of race. For more than a century, measurements of human skulls would insist that European braincases were more roomy than the rest—until more accurate testing showed all races to fall in the same range of sizes.

But the attraction of uniformitarianism, and a mass of scientific evidence that accumulated around the idea as steadily as the processes the theory predicted, won the day—so much so that, within a matter of decades, uniformitarianism had completely supplanted the theory of catastrophism. In both biology and geology, the past was seen the same way: a series of small, minute actions that produced measurable change only over deep time. Uniformitarianism was predictable, measurable, comprehensible—comfortable.

That same love of comfort often pervades economic systems. When economists look at the behavior of markets, consumers, and systems, they don't look at each event as a unique occurrence. They, quite rightly, look for guidance in past events, using the failures and successes of previous cycles to guide actions in the current situation. In doing so, economists are expressing that same connection between past and present that Lyell formulated in 1830. (Though the economists are looking through the other end of the scope—hoping to gain insight into the present by looking at the past, rather than peering at the present for clues to what went before.)

This kind of steady predictability was used as a means to replace defined benefit pensions with that most ubiquitous new financial instrument: the 401(k) plan. Within three years of its inception in 1980, more than half of American corporations were offering 401(k) plans, either as a supplement or a replacement for defined benefit pension plans. Even though the original legislation targeted executive compensation, by 1993, most corporations offered *only* these kinds of contribution-based plans for all employees. And why not? Seen over the span of decades, the value of stocks increased more quickly than the value of benefits held in pensions. Moving to contribution-based plans saved corporations billions, and employees enrolled in these plans were shown charts demonstrating that they would one day be millionaires . . . all thanks to uniformitarianism at work in the marketplace, creating a slow, steady accumulation of wealth. Though the basis of the 401(k) shift was the predictability of financial markets, the shift away from pensions itself was so abrupt as to represent an enormous "catastrophic" change. It can be argued that the whole of the economic boom experienced in the 1990s was an illusion generated by corporations removing the cost of defined pension plans from their books and placing that risk onto average workers. Debt didn't vanish, it simply moved off the corporate books.

The trouble is, they were all wrong. Lyell was wrong. Cuvier was wrong. FitzRoy was wrong. And the guy who sold you that 401(k) plan was wrong. They were all wrong because, while they stood on opposite sides of the line of uniformitarianism vs. catastrophism, they were united (to varying degrees) in a philosophy of absolutism.

The truth is that the world was formed both by gradual processes and by disasters. Anyone who adheres too strongly to either theory is likely to miss the evidence that the other is also at work. The affection for uniformitarianism among geologists became so strong that when a physicist named Luis Alvarez suggested that the extinction event at the end of the Cretaceous period might have been caused by a disaster—in particular by an object from space striking Earth—it took years before geologists would acknowledge the importance of his evidence. Biologists had become so wedded to the idea of gradual change that when Niles Eldredge and Stephen Jay Gould introduced the idea of

"punctuated equilibrium" in 1972, it was seen by some as an
attempt to return to chaotic unpredictability. (And, in fact, some
supporters of punctuated equilibrium did attempt an end run
around gradual change in a way that eventually proved to be
incorrect.) In the world of economics, market watchers were still
predicting ever-higher peaks even as the comet of the coming
disaster hung in the sky.

Lyell was absolutely right on the central point of his great
book—the same processes that shape the past also shape the
present. But the present isn't only shaped by the drip-drip-drip
of gradual change. It's also formed by asteroids, and eruptions,
and floods. The fossil record reveals that the kind of slow, steady
change that Darwin originally proposed is the source of new species, but the *rate* of change is not fixed. Not every change is predictable. Not every change happens at the same pace. And steady
contributions to a 401(k) plan are not a guarantee of eventual
wealth.

We all live in a "punctuated" world, one in which gradual processes are occasionally replaced by more momentous actions. To
Darwin's great credit, he recognized that the difference between
uniformitarianism and catastrophism was often little more than
a difference in degree. Bird wings might not appear fully formed,
but he did accept that "everyone who believes in slow and gradual
evolution will of course admit that specific changes may have
been as abrupt and as great as any single variation which we meet
with under nature."[2] Though his tendency was toward uniformitarianism, Darwin was much less rigid in applying his theory
than many of those who would come after him.

But although Darwin was willing to leaven his steadiness
with a bit of more rapid change, FitzRoy never again seemed
ready to make the move toward accepting uniformitarianism

that he had during that long voyage on the *Beagle*. After Darwin revised his volume on the natural history of the expedition in 1845, FitzRoy wanted no part of Darwin's growing insistence on deep time and gradual change. Instead, FitzRoy expressed great concern that such ideas might influence the next generation of young men and devoted himself to a defense of the formation of the world's features during Noah's flood. In 1859, when *On the Origin of Species* was published, FitzRoy felt crushed by guilt for the role he had played in helping bring Darwin's theory of natural selection to the world.

Though he was only 54 at the time, heavy responsibilities and a lifelong struggle with depression had made FitzRoy seem frail and elderly. He appeared at public meetings holding aloft a huge Bible and calling on people to believe "God, not man." For the most part, he was ignored. Even when people stopped to listen, few seemed to recognize that this was the man who had carried Darwin around the world, the inventor of the storm glass, the former governor, former MP, and vice admiral of the navy. He had become another face in the crowd, another voice lost in the storm brought about by Darwin's theory.

Perhaps that was on Robert FitzRoy's mind in 1865, when he rose from bed one morning, slipped past his wife, and took his own life with a straight razor. Or perhaps he was haunted by the fact that his family fortune was exhausted. Vice Admiral FitzRoy, always the good public servant, had taken no money for his inventions and innovations. As governor, he had made no land deals to ensure his family's wealth. As MP he had made no arrangements to secure his financial future. As harbormaster, he had taken no bribes when it came to supplying the navy. In fact, most of his money had gone for public expenses that the government was supposed to repay, but never had.

He died penniless—a personal catastrophe to be sure, and one not softened by any form of pension.

However, following his death, an Admiral FitzRoy Testimonial Fund was established to keep FitzRoy's family from poverty. Only then did the government repay the money that was owed to FitzRoy—an amount equivalent to $200,000 today. Additional funds came by way of a donation from an old friend of the family—Charles Darwin.

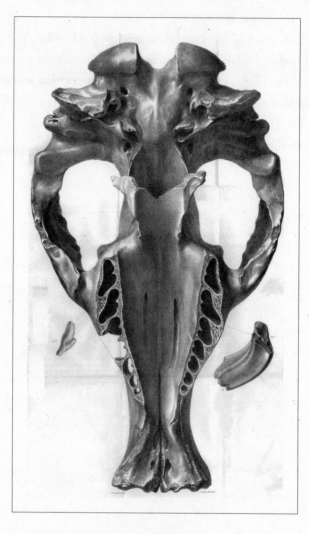

Skull of the Extinct Mammal Toxodon

FROM ZOOLOGY OF
THE VOYAGE OF HMS BEAGLE, 1838

CHAPTER 4

This Curious Subject

> It occurred to me to ask the question, why do some die
> and some live? And the answer was clearly, on the whole
> the best fitted live. . . . In this way every part of an animal's
> organization could be modified exactly as required, and
> in the very process of this modification the unmodified
> would die out, and thus the definite characters and the clear
> isolation of each new species would be explained.
>
> ALFRED RUSSEL WALLACE, *My Life* (1905)

On July 20, 1976, the Viking 1 lander separated from the remainder of the craft that had made the 60 million kilometer trip from Earth to Mars. The journey to that point had taken almost a year and had been many more years in the planning, but scientists at NASA and at the Jet Propulsion Laboratory's mission center watched nervously as the ballet of parachute and retro rockets guided the lander toward the surface. Because of the great distance involved, direct control of the craft was not possible. The scientists had to count on the probe's on-board computer—a system with only 6 kilobytes of memory, or about one-millionth that of the average PC in 2010—to hold all the commands that would bring the Viking lander to a soft touchdown.

Because of the delay caused by the speed of light over such a distance, they could not even watch events as they were happening. Everything that came across the slender radio link from Mars was a message coming out of the past. It was several minutes after the scheduled touchdown before a signal of the craft's status could be detected on Earth.

At NASA and JPL, planetary scientists held their breath. The probe—about two-thirds the size of a Volkswagen Beetle—might be resting safely on the rocky red soil of western Chryse Planitia. Or it might be nothing but a field of debris scattered in the Martian rubble. When the signal finally arrived announcing that the lander was safely on Mars, cheers (and tears) erupted across the control room. After an enormous expenditure of time, money, and supreme human effort, they were about to be rewarded with images and data from a world that, until then, had been only a blurry dot in a telescope or a few distant images snapped from orbit. Many of these scientists had worked a lifetime to arrive at this moment. They were prepared for anything. They hoped for wonders.

One hundred and forty years earlier, Charles Darwin had been the Viking 1 lander.

The *Beagle* journey was a costly scientific enterprise, five years in the execution. By the time the ship returned to England in 1836, Darwin had become a bit of a minor celebrity, especially in the clubby world of British naturalists. Darwin's winning personality and scant writings (to this point) were not the only things that endeared him to his peers. After returning from such a lengthy trip, Darwin was the possessor of something that his associates needed: data.

Heading out on the journey, Charles Darwin had been a promising student with a freshly minted bachelor's degree who had been recommended for a shipboard slot by a friendly professor. Coming back, he was the owner of notebook after notebook full of sketches showing plants no one else had seen, animals and behavior no one else had observed, and geologic formations no one else had measured. He also had tens of thousands of specimens—pickled fish, fossil skulls, and every form of plant—packed, pressed, dried, or potted. All of them were ready to be distributed to new homes. The large amount of material and information that Darwin carried would provide the grist for years of work at universities across the country. Some of his specimens would still be the objects of study almost two centuries later.

Darwin was not the only source of such data. The early decades of the nineteenth century found zoologists and botanists swamped with finds that arrived from distant corners of the European empires. Geologists were busily extending the details and distances covered by geologic maps and examining the fossils they found. There were so many basic lines of inquiry not yet opened, so many fundamental questions yet to be answered, so many opportunities for fresh discovery, that any data could easily be overlooked. Darwin's journey had given him much valuable and unique information, but with so much going on, even the Viking lander would need a little promotion to attract the attention of the nation's top scientists as they struggled through the backlog of discoveries already lined up for consideration.

Darwin enjoyed two advantages in his efforts to peddle his plants, charts, and stones. The first was a public relations campaign run by Cambridge professor John Stevens Henslow. Professor Henslow was the man who had originally recommended Darwin for the *Beagle* voyage. As professor and student, the two men had been very close. (Darwin had acquired the nickname "the man who walks with Henslow" while at Cambridge for his tendency to tag along as his professor patrolled the campus.) As Darwin transitioned from student to scientist, Henslow was there again to help him. A year before the *Beagle* returned, Henslow published a pamphlet of letters on geology that Darwin had sent from various points along the journey. Henslow then delivered copies to some of the biggest names in science. He effectively worked up anticipation of Darwin's return and built up the young naturalist into a "name" in his absence.

Darwin had another force as valuable as his portfolio of information on his side: his father. Robert Darwin might have been a doctor from Shrewsbury, but he was more than just a country physician. Robert had made careful investments that brought returns well above what he collected for house calls, and he had connections in the financial world. He brought to his son sufficient funding that Darwin did not have to immediately scramble to secure a paying position, and he helped his son gain access to offices that might otherwise have been closed to him.

Darwin had his own energy and intelligence. He had the connections opened to him by Henslow. He had the time and money provided by his father. He had a measure of public fame and sentiment provided by the romance of the *Beagle* voyage. For any scientist at any time, it was a welcome combination.

And then, 23 years later, he wrote *On the Origin of Species*.

At least, that's the way his life is portrayed in most text books—go on *Beagle*, 1832 to 1836; write *On Origin of Species*, 1859. Everything in between is just 20 years of Darwin warming up to his subject.

It's true enough that Darwin took some time to put together the ideas that would form natural selection, but he did a lot more over those two decades than ponder finches. First there were the *Beagle* volumes to be written. Darwin eagerly agreed to unrealistic timelines that forced him to write at a blistering pace and work ever longer hours as the deadline loomed (a phenomenon that remains decidedly unevolved in the life of many authors). At the same time, he was making rounds of museums and universities, following up on Henslow's public relations work with personal meetings.

During this time Darwin first met many of the scientists who had been crucial to the advancement of geology and biology over the previous decades and many who would be equally funda-mental in the debate to come. Charles Lyell, whose book had been instrumental in shaping Darwin's work aboard the *Beagle*, was eager to talk about the geology notes. Lyell then introduced Darwin to other experts who could help unscramble the nature of the specimens Darwin had ferried home from South America.

Even though Darwin had been there to collect the specimens and make the observations, his publications on the voyage would be only a minor part of the ultimate picture of the data collected by the *Beagle*. Just as the Viking data would be examined for decades after the craft's arrival on that dusty plain, the *Beagle* data would prove revelatory. At the Royal College of Surgeons, Richard Owen would discover that among the fossils were extinct giant sloths and an extinct rodent the size of a hippopotamus. Ornithologist John Gould of the Zoological Society of London began unraveling the relationship of the famous finches, which Darwin had not even realized were finches and had thought to be members of multiple families. A five-part series on the zoology of the *Beagle* specimens would be edited by Darwin and would feature not only Owen and Gould but other experts on reptiles, fish, and mammals as well.

By 1837, Darwin was engaged in his editorial duties, in writing his own papers, and in seeing the specimens from the *Beagle* distributed. He was also starting to put together the first rough sketch of his own idea on the transition between one species and another. In the spring of that year, Darwin doodled a branching tree in one of his notebooks (see Figure 4.1).

In his published journal of the *Beagle* journey, Darwin skated around the issue of evolution. He noted that the animals living in the Galápagos were similar to the species found in South America but didn't extend this thought to suggesting that the Galápagos animals had ancestors on the mainland. Instead he made an oblique reference to the "centers of creation" idea put forward by Lyell and stopped there, saying, "there is not space in this work, to enter into this curious subject."[1]

Darwin then spent the better part of a year being prodigiously ill with an undiagnosed disease that may have hitched a ride back with him across the oceans. His illness might also have been something acquired from the none-too-sanitary water supply of London in the 1830s (a period when 1,000 Londoners a month were dying of cholera alone), or it may have been brought on by the stress of juggling so many responsibilities. Whatever the cause, the illness would strike Darwin frequently throughout the rest of his life and lead him to seek treatment from a number of doctors.

As he recovered from this first bout with illness, Darwin tumbled into a romance with his cousin, Emma Wedgwood. She was quick-witted, musically talented, athletic, adventurous, and had a reputation for not paying particular attention to either fashion or housekeeping. She was charming. She had been forced to refuse several previous offers of marriage while nursing her sick sister and mother, but her sister had recently died. Emma

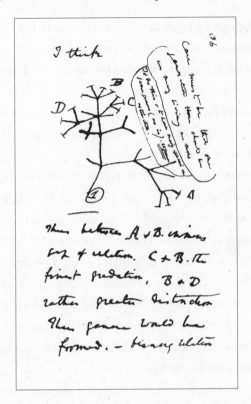

FIGURE 4.1. DARWIN'S INITIAL SKETCH
OF BRANCHING EVOLUTION, 1837

was for the first time able to think of her own future. Darwin fell for her immediately.

By then his ideas about the transmutation of species had become clearer, and he was certain that he had a major contribution to make to the understanding of this concept. He was also certain that his ideas would bring both attention and trouble, and felt that he should warn Emma of where his research was taking him. For her part, Emma held deep personal beliefs about religion. As Darwin suspected, his thoughts troubled her. But what bothered Emma wasn't so much the biological details of

Darwin's ideas. Instead, she voiced a concern that had already been leveled against Lamarck and others who advanced theories that appeared to sever connections between the visible world and the spiritual world—such ideas could lead to abject materialism. What worried Emma wasn't the status of Adam and Eve, it was whether a world explainable in purely physical terms might be a world without a moral foundation.

Conversely, Darwin's growing doubts about religion were not based on his own discoveries. Like Lyell, Darwin was able to separate his scientific ideas from his more spiritual thoughts. It wasn't evolution that caused Darwin to declare himself an agnostic, it was damnation. In particular, he could not abide the idea that people could be damned for all eternity—some of them for doing nothing more than having had the bad luck to be born in a place where they never encountered the Christian faith. Darwin and Emma discussed, debated, and worried that issues of religion would come between them.

This point of contention might have become an impassable roadblock, but Emma's own beliefs weren't based on strict interpretation of scripture. Instead she had spent long periods studying and asking her own questions before forming her beliefs. She didn't begrudge Charles his doubts or question his intentions—she believed in questioning and thinking through issues even when they went against the grain of society—but she wasn't persuaded that Darwin had found the right answer. And she worried that an issue of such importance would always be between them.

> When I am with you I think all melancholy thoughts keep out
> of my head but since you are gone some sad ones have forced
> themselves in, of fear that our opinions on the most important
> subject should differ widely. My reason tells me that honest &

conscientious doubts cannot be a sin, but I feel it would be a painful void between us.[2]

Darwin fretted over the same issue. He was increasingly consumed with the idea of what was then still generally known as "transmutation," and had taken to pestering everyone from professors to pig farmers to expand his insights. Having fallen in love with Emma, he worried that he couldn't help but hurt her.

In the end, they each found the other's charms exceeded the challenges. In November of 1838, Emma accepted Darwin's proposal. She was 30. He was 29. Shortly afterward, Emma would reassure Darwin that she knew he was sincerely seeking the truth, and that such a search "could not be wrong."[3] Still, Darwin would do his best not to phrase his ideas in ways that Emma might find offensive.

It was a bit of a late start for both of them. Emma and her "dear Charley" launched almost immediately into having children. As their family began to grow, so did Darwin's ideas on evolution. He pored over his notes from the *Beagle*, and cross-referenced his observations with those of others aboard.

It was FitzRoy who had actually spotted the location of the various finches that would eventually become so identified with Darwin's theory. Darwin had paid little attention to the birds while actually in the Galápagos. In fact, FitzRoy also pointed out that the tortoises in the Galápagos varied from island to island—something Darwin had overlooked. At that point on the journey—after the *Beagle* left South America and was at the start of its long journey back to England—Darwin's letters seem to indicate that he was distracted and possibly suffering a severe case of home sickness.

At home in London, Darwin finished off his editorial work

on the *Beagle* volumes and then turned straight away to addressing one of the most controversial scientific topics of the day—how
coral reefs and atolls were formed. A century and a half later the
knowledge of how reefs form may not seem important, but for a
nation that depended on a great navy composed of wooden-hulled
ships, coral reefs were a serious business, and how they formed
was a momentous concern. Darwin's book *The Structure and
Distribution of Coral Reefs* was nearly as controversial in its day as
Origin would be two decades later. Darwin's theory on atolls ran
directly counter to the prevailing theories of the day, and Darwin
was still addressing opponents of his atoll ideas years after natural
selection became the focus of public debate. As it turned out,
Darwin was right. As usual, he had based his theory on direct
observation and measurement, as well as careful reasoning. He
would later be hired to write a book on geology for the navy, to
aid officers in identifying geological formations of all types.

Darwin continued to work through his ideas on the transmutation of species and continued to ruminate the theory around
with friends, but even then he had to be cautious. Resistance to
the idea of evolution among scientists at that time had more to do
with the large number of half-baked evolutionary theories that
had been published than with stiff religious thinking. Lyell and
Owen, who had become both friends and collaborators with Darwin, resisted the notion of evolution. It was one thing for Darwin
to engage in "radical debate" among the free-thinking students
and professors at the University of Edinburgh, but now that he
was working with the more reserved professors of Cambridge
and acting as secretary of the Geological Society, he took pains to
make his arguments as close to perfect as he could. This attempt
to clarify his ideas before making them public, combined with
his increasing bouts of ill health, made him fear that he might not

live long enough to actually see his ideas on selection published. The thought troubled him enough that he produced an abbreviated 230-page "memo" describing natural selection, and he asked Emma to see that it was published if he didn't survive to produce a more elaborate draft.

Even after Darwin and Emma took their growing family to a handsome new home outside the city, the amount of work to be done slowed only a little. One of their children died in infancy—a tragedy all too common in the 1840s—but the family continued to expand. Darwin wrote two books on geology in the space of two years. Emma gave birth to five children in six years. The house was filled with children, books, visitors, and experiments. Their home was nearly constantly in the midst of some kind of expansion or revision.

He might well have published his ideas on evolution in the 1840s, but for one thing: his ideas had competition. Not only did Lamarck's theory still circulate in several forms, but for several years there had been a growing movement toward what was called *natural theology,* and in particular toward "progressive creationism." This latter theory accounted for the transition of species by a kind of *rolling creation* in which God had not populated the world over one seven-day period, but was constantly engaged in introducing new forms. This explanation solved the problem of extinction because holes in nature were filled with new forms. The theory also explained the issue of adaptation, as new creatures were created with place and purpose in mind.

In 1844, a book called *Vestiges of the Natural History of Creation* had been published anonymously by a Scottish journalist, Robert Chambers. *Vestiges* united natural theology with many of the popular ideas about evolution at that time and put those ideas into a context that the general public could understand.

Chambers's version of evolution, like Lamarck's, had a sense of direction and of there being some "more perfected" forms.

> Not one species of any creature which flourished before the tertiary . . . now exists; and of the mammalia which arose during that series, many forms are altogether gone, while of others we have now only kindred species. Thus to find not only frequent additions to the previous existing forms, but frequent withdrawals of forms which had apparently become inappropriate—a constant shifting as well as advance. . . .[4]

Most shockingly for the reading public, Chambers, having built his argument around the idea of a world propelled toward improvement and better alignment with a divine plan, then suggested that God was not required to intervene at every step of this plan. The idea that God was reaching down to personally create each new worm and beetle was "surely . . . too ridiculous to be for a moment entertained."[5] However, Chambers didn't then offer any sort of scientific system by which evolution might occur. Instead, he offered evolution as a kind of further expression of God's power. God didn't just create static species; He created species that had within them the seeds of a new creation. It was as if God were selling bicycles that contained within them the plans for jumbo jets.

By offering a book that expressed evolution as an extension of the divine, Chambers hit upon the formula for an instant bestseller. His book raced through numerous printings and multiple editions, during which the author provided new examples and illustrations. The book would remain popular for more than a decade. But Chambers's understanding of the science involved was far from complete, and because his book failed to offer any natural force driving evolution, the work was instantly dismissed by scientists. For the most part they felt that *Vestiges* only confused the issue by pretending to offer a solution but proving no mechanism.

Darwin realized that if his work was going to be taken seriously, he would need to prepare for a barrage of objections. He wanted to avoid the scorn his friends directed toward works such as *Vestiges of Creation*. He began to expand his own work, trying to anticipate the arguments his opponents would muster. In the meantime, he did his work for the navy and began a series of essays on barnacles. Barnacles, which spend most of their life stuck in the same place, may not seem the most interesting of subjects, but Darwin was interested in what these stripped-down animals held in common with larger creatures. Darwin would eventually devote his last book to the lowly earthworm. He seemed to have a particular fondness for things that were overlooked.

In 1849, a bout with scarlet fever went through the household, which then included six children, evenly divided between boys and girls. All three of the girls came down with the disease, but it seemed to strike Anne, the oldest, more severely than the others. At eight years old, Anne was—and Darwin did not deny it—her father's favorite. Compulsively neat when both her parents were not, Anne was forever tidying up after her father or smoothing down his unruly hair. She liked to help Darwin with his experiments, to offer her own ideas, or simply to be near her father when he was working. On occasion she and her brother William acted as an audience for Darwin as he settled on words to explain the idea that was continuously growing in his mind.

The two other girls soon recovered from the fever, but from that point on, Anne's health was delicate, and she began a slow decline. Most historians believe the girl had contracted tuberculosis. In 1851, Anne's situation went steeply downhill. Several times she seemed on the brink of death, and each time her recovery was less complete. By midyear, Anne's situation was dire and Darwin was desperate. He bundled the child up for a trip to the springs at Great Malvern, whose waters had seemed helpful when Darwin himself was suffering from one of his own rounds of illness a few years earlier. It didn't work. She died there, and a devastated father returned home to his reduced household.

> Oh that she could now know how deeply, how tenderly we do still and shall ever love her dear joyous face.[6]

Recovering from Anne's death took longer than recovering from his own illness, particularly because Darwin's mourning was overlain with a thick layer of guilt. At the time, the germ theory of disease was not by any means popular. Most doctors believed that diseases arose primarily out of "miasmas"—noxious vapors— and how they were communicated from one person to another was a mystery. Darwin had another theory. Having spent years focusing on the inheritance of characteristics between generations of animals, he worried that Anne's sickness was related to his own. While talking with breeders of domesticated animals from pigeons to cows, he had heard many times how mating closely related animals over generations could lead to sickliness while crossing distantly related animals generated particular vigor. Darwin had married his first cousin, a common enough practice at the time. But he ached with the thought that his little girl might have died because of some intrinsic flaw that he had passed along to her or because he and Emma were too closely related. He worried that his remaining children might suffer because of his choices.

Gradually Darwin returned to his writing and his research,

but his health continued to be a problem. In 1853, the long hours of work spent studying simple barnacles earned him recognition by the Royal Society. By 1854 it was clear to Darwin that multiple volumes would be needed to lay out his arguments and defend his positions on natural selection. He fretted, as he had a decade before in the midst of illness, about whether he would live long enough to see the work published. With the assistance of his son William, he carried out new experiments designed to answer possible arguments against his work.

Key to Darwin's argument that animals had developed, spread, and adapted to different environments was the idea that they could *get* to those environments, including such distant locations as the Galápagos or the even more isolated Easter Island. If these islands had not been directly populated by providence, the inhabitants must have reached these lands under their own power. For birds, such migration was easy enough to believe. Even for turtles or lizards, a few days of paddling and a favorable current might transport a creature from one island to the next. But what about plants? Some plants had seeds small enough to drift on the winds, but many did not. A palm tree could not exactly lift itself from one place and swim to the next. How could so many varieties of plants cross such vast stretches of salt water when they were incapable of movement?

With the help of his son William and a bathtub, Darwin conducted a simple experiment to find out. He reasoned that many seeds must end up washed into rivers and from there into the sea. All that was required for a plant to make a sea crossing was that its seeds be capable of floating for an extended time and still be viable when they washed onto some distant shore. To test this, he filled the tub with salt water and dropped in seeds of various types. Each day Willy joined his dad to check to see which seeds were still floating, and to test whether those that had sunk to the bottom were still capable of sprouting. The experiment was a success: most plant seeds were more than capable of undertaking the required journeys.

Gradually, Darwin began to share more of his thoughts with Lyell and with his best friend, botanist Joseph Hooker. Hooker was increasingly convinced that species did shift. Lyell was not. Even less sold on the idea than Lyell was the man who might as well have been Viking 2—Thomas Huxley.

A self-taught anatomist from impoverished beginnings, Huxley had worked hard to get the educational opportunities that had come so easily for Darwin. He had taken a spot on board a scientific expedition—not as a gentleman on a sort of post-graduate valedictory lap, but as a surgeon's mate in order to collect meager navy wages. From aboard the HMS *Rattlesnake*, Huxley had sent back his own scientific papers, and by the time he returned to England he was well regarded as a naturalist. At age 26, he had not only been elected to the Royal Society but awarded one of its highest honors. Overall, his education and travel experience closely matched Darwin's—and Viking 2 was not buying evolution. His thoughts on both Lamarck and on *Vestiges of Creation* were sharp enough to cut and sharp enough to keep Darwin honing his arguments.

But it wasn't the opposition of Viking 2 that would eventually kick Viking 1 into high gear. While Darwin was still working his experiments and expanding his evidence, his old friend Charles Lyell noticed a paper in the *Annals of Natural History* that had some interesting ideas closely paralleling those he had heard from Darwin as far back as 1842. The paper was not from a familiar name, but from a young man then sweating through a tropical fever in Malaysia: Alfred Russel Wallace. When Lyell and Darwin next talked, the older man told Darwin about the paper and warned him that this Wallace seemed to be coming very close to the ideas that Darwin had developed. Darwin was not concerned. From Lyell's description, he took Wallace to be another advocate of the same kind of natural theology put forth in *Vestiges of Creation.*

But the talk with Lyell nagged at Darwin. He started sketching out a short paper for publication in the near future, and accelerated his contacts with biologists as he collected the information he would need for his planned multi-volume work. Among those he contacted for information this time was Wallace. In response, Darwin not only got the data he requested, but a letter from Wallace asking about Darwin's work. Darwin was pleased to see that Wallace was working on problems of distribution—the subject that had kept Darwin's tub full of seeds (and occasionally eggs) for a year—but he seemed to still believe that Wallace was basing his studies on the progressive creationism path. In his response, Darwin wrote that he was planning to publish a book in the next few years and would be going "much further than" Wallace.[7]

In 1858, when the 49-year-old Darwin was halfway through his book, he received a second note from the 35-year-old Wallace. Darwin was shocked to see that the note contained the theory that he had been working on since 1837. Wallace's paper was short and free of all the "defense" that Darwin felt compelled to provide; it also used different terminology—the phrase "natural selection" did not appear. But there was little doubt that his conclusions were the same as those Darwin had drawn, and his arguments very similar to the ones that Darwin employed. Suddenly Darwin had a new fear that supplanted his concerns of being taken seriously. He might be bypassed entirely and left on the sidelines of the discussion.

However, Darwin made no attempt to hide Wallace's work. He passed it on to Lyell, with a note saying that he thought Wallace's paper excellent, and he could not have done a better job of summing up the topic. Wallace had not asked for publication, but Darwin offered to send the paper to any publication that Wallace might approve.

At that point, Robert Brown died.

Brown, an accomplished botanist and skilled hand with a microscope, had been the first to name the nucleus of the cell and the first to describe the constant random jittering of very tiny structures (still known today as *Brownian motion*). He was the first keeper of the Botanical Department at the British Museum of Natural History, the first botanist to explore areas of Western Australia, and the first to describe over 1500 plant species. Son and grandson of Scottish ministers, army veteran and former ship's surgeon, he was friend or mentor to many of the great scientific minds of his age. He was also 84, so his death in the spring of 1858 was sad, but not exactly shocking.

However, his loss was also both an inconvenience and an opportunity. Brown was the sitting vice president of the Linnean Society, the oldest organization dedicated to the natural sciences. The botanist's funeral had overlapped with the normal date of the society's June meeting, and the next regular meeting was not scheduled until November. It was thought the venerable organization could not go so long with the post of vice president vacant, so Brown's death led to the establishment of a special session to deal with the election of a replacement. It was that meeting that opened the door.

An additional meeting meant an additional chance to submit papers. Presenting a paper at a Linnean Society meeting was the first step to having that paper published in one of its journals. For a naturalist seeking respectability, the *Journal of the Proceedings of the Linnean Society* was the closest thing to a modern peer-reviewed journal available in its day and was one of the two most respected journals in the field.

Darwin's friends Lyell and Hooker arranged for a joint presentation of Wallace and Darwin's work at this hastily arranged meeting. Because the meeting was only a few weeks away, there was little time to harmonize the writing. The paper consisted of two partial chapters from two separate works, neither of which had been altered to smooth the text into any semblance of a single work. In addition, the paper included letters of introduction from Lyell and Hooker and a bit of Darwin's personal correspondence dating back almost two decades that was inserted as proof of precedence (in other words, as proof that Darwin wasn't merely plagiarizing Wallace). The whole production was, by any standard, a bit of a mess.

The situation of the authors was equally messy. When the meeting began on July 1, 1858, Wallace was still in Malaysia. There was no question of his being able to return in time to attend the meeting. When Lyell and Hooker set it up, they assumed that Darwin would handle the presentation. That wasn't going to happen. For Darwin, June and July of 1858 had turned into a waking nightmare. Scarlet fever was back.

After Anne's death in the spring of 1851, Charles and Emma Darwin had two more children. The first was born only a month after Anne died. The last of their children, Charles Warring, was born at the end of 1856 when Charles Darwin was 47 and Emma was 48. The little boy was slow to walk and slow to talk. Some suggested that Charles Warring might not be developing normally. Darwin, who worried incessantly about what his illness and his close relationship to Emma meant for his children, defended his infant son fiercely, calling him "intelligent and observant." It's quite possible, considering the age of his parents, that Charles Warring was born with Down syndrome.

The diagnosis of Down syndrome may be conjecture, but certainly the 18-month-old child contracted scarlet fever as it ran through the area around Darwin's home in June of 1858. Other members of the household were also ill. While Lyell and Hooker scrambled to get the paper on the Linnean Society's schedule, Darwin was busy nursing his children. Three days before the meeting, little Charles Warring died. Darwin was devastated; he would be attending no meeting.

Most of the Linnean Society meeting on July 1, 1858, was taken up with a long eulogy to the departed Dr. Brown, recalling his 60 years of involvement with the society. More time was taken with the election and with the various items of business on the society's agenda. In those days, the society shared tight quarters with other organizations at Burlington House. (The much more spacious accommodations the society enjoys today were built a decade later.) Those gathered for the meeting had already been packed into the warm meeting hall for some hours before the reading of the papers began. Papers presented to the society were usually read aloud, in full, and by the authors when possible. Those attending the meeting that evening found themselves sitting through papers on the botany of Angola, the habits of lampreys, worms that burrowed into clam shells, and a newly discovered form of extinct trilobite. A total of six papers were read—making it an extremely full meeting.

When it came time for the presentation of the Darwin-Wallace paper, the task of reading the text fell to George Busk, the under-secretary. A professor, physician, and retired naval surgeon, Busk had at least a nodding acquaintance with Darwin (they would meet again years later, when Busk treated Darwin for his ongoing ailments.) He also had at least a passing familiarity with the subject of the text. He may have asked to be the one to read this particular paper.

Hooker would later report that the reading was received enthusiastically, but this report was definitely of the minority view. The sheer length of the meeting—more than five hours—and the number of papers delivered may have been enough to drown any interest or discussion. Though the paper was among a handful selected to be printed in one of the society's journals, even that publication seemed to garner little interest.

At the end of 1858, the president of the Linnean Society lamented that the year had not "been marked by any of those striking discoveries which at once revolutionize . . . science."[8] He could not have been more wrong.

*Alfred Russel Wallace
in 1908*

LINNEAN SOCIETY

*Charles Darwin
in 1881*

PHOTOGRAPH BY
HERBERT ROSE BARRAUD

CHAPTER 5

Sometimes They Even Talk Alike

How extremely stupid not to have thought of that!
THOMAS HUXLEY (1858)

In 2008, the Toyota Prius was the best-selling hybrid car in the world. In 2009, Honda introduced the Insight, its new hybrid model, which looked . . . a lot like the Toyota Prius. The reason for this similarity wasn't some plot among Honda designers to confuse their product with the Toyota—it was merely a case of form following function. Both automakers were trying to craft a vehicle that maximized available interior space and provided the best gas mileage possible. To reach that second goal, they needed to minimize the drag caused by air flowing over the car. The result was two vehicles that, if not identical, could pass for siblings.

The same rules hold true in many areas of commerce. An Apple iPhone looks a lot like an HTC Touch, which looks a lot like a BlackBerry Storm because all of them are trying to give their users the largest area of screen possible while keeping the shape and size of the device suitable for the average pocket. From bicycles to washing machines, few pieces of technology really stand out from the crowd. Chances are that two items designed for the same task will have very similar forms.

A woman named Mary Anning was one of the first to discover that what applies to gadgets today applied to animals through the ages. Mary's father died of tuberculosis in 1810, leaving 10-year-old Mary and her 12-year-old brother, Joseph, to provide for the family. They made a living by scouring the cliffs and beaches in Dorset, looking for fossils that weathered out of the limestone in an area known to collectors as the "Jurassic Coast." Collecting fossils had become fashionable, and while the specimens she found were at first barely enough to keep the family fed, Mary soon gained a reputation for turning up extraordinary finds. At the age of 12, she found the first complete skeleton of an animal called an *ichthyosaur.*

Ichthyosaurs were streamlined creatures with long snouts, large eyes, and powerful tails. Later fossils would show that they had large dorsal fins on their backs. They sped through the seas of the Jurassic (and Triassic, and Cretaceous) racing after fish and shellfish. If one were to pop up off the Dorset shore today, observers on land might well believe that they were watching a dolphin as the animal dashed and jumped among the waves. But ichthyosaurs were not related to dolphins. They were reptiles, a family that originated from lizard-like ancestors instead of the land-dwelling mammals that would eventually produce dolphins and whales.

The reason that ichthyosaurs look much like dolphins is

because both are performing the same role in a similar environment. Both are predators that hunt their prey in the ocean. The long snout filled with banks of sharp, narrow teeth is an excellent mechanism for catching fish. The large eyes developed because sea water is clear and sight is an important aspect of tracking and capturing fast-moving fish in shallow water. (Those few species of dolphins that today live in the muddy waters of rivers are either nearly or completely blind.) The overall shapes of both animals were similar because both faced the same governing requirement—moving quickly and efficiently through water.

The ease with which an object can move through a medium such as air or water is measured by looking at the drag coefficient. A high drag coefficient means that an object is not very efficient at moving through the medium and will need to expend more energy. A person walking in air has a drag of about 1.2 (which is not particularly good). Older cars have drag coefficients as high as 0.7. The Toyota Prius and the Honda Insight both have drag coefficients below 0.3. That's not bad, but jet aircraft can do as much as 15 times better. The most slippery supersonic planes have a drag of less than 0.02.

Many land animals (people included) get along with shapes that are not particularly low in drag, but water is 830 times as dense as air and 50 times as viscous. The penalty for a water-going creature that isn't shaped to avoid drag is many times higher than it would be if the same creature were moving in air. The drag coefficient for dolphins is less than 0.004—much better than the sleekest aircraft. Ichthyosaurs looked a lot like dolphins (or dolphins look a lot like ichthyosaurs) because that shape is a pretty good design for moving in water. Both animals were refined by the same powerful forces: the need to defeat drag and the force first explained by Charles Darwin and Alfred Russel Wallace.

Two types of organisms moving toward similar solutions to the same problem, despite having very different origins, is common in fossil records (and in the living world around us). Bats and pterosaurs had very different ancestors but made some of the same adaptations to flying. Marsupial animals in Australia developed many of the same solutions as their placental relatives in the rest of the world, leading to such informal designations of some Australian natives as the "marsupial mouse," "marsupial cat," or "marsupial wolf."

Both creatures and consumer products adapt to their "environments." For creatures, this environment can encompass something as simple as the medium through which they move, but it's also the food they search for, the predators that hunt them, the temperature of the water, the chemistry of the air, their competitors, their diseases—everything around them. Likewise, products exist in an environment that includes consumers, competitors, retail markets, economic conditions, resource availability, and items as ephemeral as "style."

Ideas also form in an environment. They require a foundation of previous ideas, the availability of raw information, and a society that supports new thinking. In 1684, the German mathematician Gottfried Leibniz published the first paper based on the calculus he had invented over the previous decade. Isaac Newton didn't publish his first work of calculus until 1693 but said he had been working on his version since 1666. For three centuries, adherents of both men have argued over who really invented calculus, but the answer is both. On the technological front, the list of near-simultaneous discoveries is lengthy. Was the first practical electric light bulb invented by Thomas Edison or by Joseph Swan? Did Alexander Graham Bell invent the telephone, or was it Elisha Gray? The radio, the television, and the medical MRI all have multiple claimants to the title of "inventor." Even when it comes to relativity, several others anticipated at least some of what Einstein would so brilliantly assemble.

When Charles Darwin opened a note from Alfred Russel Wallace in the spring of 1858 and found that Wallace had deduced the same ideas on natural selection that Darwin had been thinking over since 1837, the simultaneity wasn't too surprising. Both men had been inspired by reading the Reverend Thomas Malthus, whose 1798 essay fretted over the possibility of the human population outracing its food supply:

> The power of population is indefinitely greater than the
> power in the earth to produce subsistence for man. Population,
> when unchecked, increases in a geometrical ratio. Subsistence
> increases only in an arithmetical ratio. A slight acquaintance
> with numbers will show the immensity of the first power in
> comparison with the second.[1]

Malthus limited his speculations to man, but both Darwin and Wallace began to think about what the limits of population meant for other species. Any creature that expanded its population generation by generation must inevitably run into limits. Limits on food. Limits on space. Limits on the proper kind of habitat. In fact, most creatures bump against these limits all the time. The environment supports as many individuals as possible.

Both Darwin and Wallace also knew that differences exist between individuals. These differences were evident in the observations of animals and plants they had made on their travels, and equally obvious from the domestic animals they saw all around them. Imagine Victorian England: today, the idea that the streets of London were full of horses seems odd and somewhat charming—so long as you're not the one in charge of cleaning the streets. But those streets were also full of sheep, goats, cows, chickens, and other animals. Even people who didn't live on farms were constantly exposed to domesticated animals of all types. Their availability is one reason why both Darwin and Wallace made extensive use of these animals as examples in their work.

And those are all the ingredients needed for natural selection. Environmental pressures limit which organisms are able to hang around and reproduce. Not all organisms of the same species are identical. Those that live, love, and have offspring tend to pass their traits to the next generation.

Populations of organisms can, and do, contain many variations all the time. Often the variations they carry don't have any effect on their ability to produce offspring. Is a blue-eyed wolf more likely to have pups than a gray-eyed wolf? Only if the wolf's potential partners have a thing for a certain eye color. When conditions are good, many traits have such a minimal effect on reproduction that they linger in the populace. What does it matter if you're able to skate by on less food when food is abundant? It doesn't, until it's not.

A change in any aspect of the environment can suddenly turn what had been just another variant into either an advantage or a detriment. Being able to survive on less water might not be valued, until there's a prolonged drought. Tolerating heat or cold better than your neighbors might have no value, until extreme conditions strike. Being the slowest rabbit in the area might not matter, until foxes arrive. Big can be big when there's plenty of food, but shortages may give small the edge.

Differences exist among individuals. Some of those differences tend to be inherited by the offspring produced by those individuals. If a selective pressure acts against those differences, you get evolution. The end. In any given generation, the differences may be small. The edge given to one variant over another can be razor thin. Still, evolution moves on, aided by that secret ingredient spelled out by James Hutton—long periods of time.

The simplicity of the idea is part of what made it hard for some people to accept. Some scientists among Darwin's immediate circle of friends knew that evolution had happened, but they thought Darwin's mechanism was just not sufficient. There had to be more to it. Over the next 60 years, natural selection would come in and out of favor as the driving system behind evolution, and that bald simplicity would be one of the biggest problems. Even in Darwin's time, his most vocal opponents were not those who didn't believe that evolution had taken place. He faced more opposition from those—such as anatomist Richard Owen, who had worked with Darwin on specimens returned from the *Beagle*—who didn't believe that Darwin's simple process could be responsible for all the variety in the world.

However, this same simplicity was also one of the theory's biggest assets. Once explained—and Darwin couched the idea expertly with many examples and answers to the questions he anticipated from opponents—there was a forehead smack heard round the world. Many who had been followers of some earlier system, or doubters of evolution altogether, saw the unmistakable nature of the idea and became vocal supporters.

Thomas Huxley's reaction was probably typical of many who

read *On the Origin of Species* when it reached the stands in 1859. On reading *Origin*, Huxley noted that he was "extremely stupid not to have thought of that!"[2]

Huxley could have thought of it. So could any one of hundreds of naturalists who shared similar backgrounds, information, and experience. The many examples of animals and plants that Wallace had cataloged in his Malaysian wanderings and Darwin had seen in his shipboard travels gave them both an important nudge toward thoughts of natural selection. But others had seen as much and done as much. Certainly men like Huxley, who had also been part of a round-the-world scientific voyage and had access to the same data that Darwin had used, or men like Owen, who had examined and cataloged tens of thousands of creatures both fossil and living, could have come up with the same idea.

That these other men did not arrive at the same conclusion shows only how hard simple things can be. Natural selection looks simple, and it is simple, but out of that simplicity endless complexity can be generated. Looking at the world, most scientists (and nonscientists) saw only the complexity. The living world was so massive, so intricately interwoven, and so complex, that people looked for massive, intricate, complex answers. The real insight that Darwin and Wallace shared was that complexity could spring from simplicity. There was no need to go looking for a complex solution. There was no need for divine intervention at the appearance of every species and no need for a mechanism that propelled the world toward some hidden plan.

In fact, for all their apparent similarities, Darwin and Wallace reached their similar conclusions by different means. Darwin, despite originating an idea that's among the most powerful in all of science, was something of a conservative thinker. The words of his textbooks and professors put him on a path where he gradually, almost tentatively, moved toward the idea of natural selection. He then spent decades expanding, testing, and validating his thoughts. Wallace was the real radical, a man who welcomed odd thoughts and unorthodox notions. Wallace would go on not only to elaborate on natural selection, but also to note that many animals employed "warning colors" to announce that they carried poison or some similar threat. He would be the father of biogeography, doing much more than Darwin to explain and establish the distribution of plants and animals. Wallace's ideas were essential to explaining how the accumulation of small changes in natural selection led to the development of separate species.

Wallace's nimble mind was also prone to making less sensible jumps. He would become an avid spiritualist who attended séances and supported channelers. (On one occasion, Wallace was a character witness for a would-be spirit medium accused of fraud, while Darwin helped to fund the prosecution.) Wallace was something of a scientific James Dean—traditional thought prompted him to rebellion.

Both men were brilliant. Both were capable of highly original thinking. But Darwin was persistent, dogged, and obsessive when it came to working out every implication of an issue. Wallace was more mercurial, always ready to tackle something new. The two men came together on natural selection because they were driven there by the data. Once glimpsed, the idea of natural selection was as powerful and necessary for them both as the streamlined bodies of the dolphin and the ichthyosaur. Some ideas are so compelling that they shape the direction of the whole world.

Some people are equally compelling. A decade before Wallace's letter arrived in Darwin's mailbox, the Geological Society realized that they had to do something to recognize one of the most important paleontologists in the nation—Mary Anning. The woman who had started as a child by selling fossils to support her family continued in this task all her life. The beautiful specimens she recovered, among the most complete ever found, provided additional proof of extinction and showed how different the seas of the Jurassic had been. She uncovered not only the ichthyosaur but multiple forms of the long-necked plesiosaurs, extinct fish, crocodiles, and even a delicate pterosaur that had been unlucky enough to plunge into the ancient sea.

Anning sold her discoveries to men like Richard Owen, who described them and gained fame in the process, but she was never more than a step away from poverty herself. Still, the importance of what Anning had done and the dedication with which she pulled out one complete skeleton after another won her admiration among the men who were rewriting science. In 1847 the Geological Society was, like all such societies at the time, restricted to men only, but by overwhelming vote the society made Mary Anning an honorary member. Only a few months later, at the age of 47, she died of breast cancer. On her death, a stained glass window was erected in the church of St. Michael the Archangel. The inscription on the window reads "This window is sacred to the memory of Mary Anning of this parish, who died 9 March AD 1847, and is erected by the vicar and some members of the Geological Society of London in commemoration of her usefulness in furthering the science of geology, as also of her benevolence of heart and integrity of life."

The window shows how valued Anning's contributions were, but another remembrance of Mary Anning is perhaps an even better claim to fame—though not quite as grand as a stained glass window. Mary Anning, the child who sold fossils on the beaches of Dorset to save her impoverished family, is said to be the inspiration behind the old tongue-twister "She sells sea shells by the sea shore."

Memorial to Those Who Passed Through the Workhouses, West Midlands, England

PHOTOGRAPH BY TONY HISGETT, 2009

CHAPTER 6

The Time Machine

So, in the end, above ground you must have the Haves,
pursuing pleasure and comfort and beauty, and below
ground the Have-nots, the workers getting continually
adapted to the conditions of their labor.

H. G. WELLS, *The Time Machine* (1895)

On a warm summer evening, a small group of children gather in
a grassy space between their homes. Under the combined light
of street lamps and stars, they hold hands and begin to circle
slowly, sneakers and sandals squeaking in the dew. Gradually
the circle picks up speed, until the smaller children are nearly
flying, their hands held tight on either side, their legs swinging
above the ground. As the rotation grows faster and faster, all of
them begin to sing.

> Ring o' ring o' rosies,
> Pocket full of posies,
> Ashes, ashes,
> We all fall down.

At the moment they reach the final words, the children release their grip on one another and the circle flies apart. They tumble to the ground, dizzy and laughing. Quite quickly the children climb back to their feet and brush away the damp grass from their legs. They come together to start the ritual again, and again they sing their brief song, all the while completely unaware of the dark story behind their innocent play.

Embedded in the simple nursery rhyme are lingering terms from one of the darkest episodes in the history of Western Europe. The "ring o' rosies" recalls the reddish marks that broke out on the bodies of people infected with bubonic plague. The "pocket full of posies" was a common ward against the disease, and the "we all fall down" evokes the sudden demise that overcame so many during the Black Death of the fourteenth century.

This very satisfying bit of knowledge fills in the history behind a common activity, matches details against aspects of the Black Death that may be vaguely recalled from a book or history class, and makes a shivery association between the innocuous activity of children and the horrible fate that overwhelmed a third of Europe. It's a good story in every way—except that it is not true. The nursery rhyme actually didn't appear until the nineteenth century, more than 500 years too late to mark the passage of the plague. Posies were *not* carried to ward off the Black Death. Most of the words in the song are simple nonsense rhymes that have been altered and altered again as each generation of children has handed them to the next. Almost as many variations exist as children to sing the song, and none of them have anything to do with plague.

Still, it *is* a very nice story, and as with a thousand other myths of our age (pop rocks and soda, anyone?) a little thing like not being true won't stop the dark and secret history of ring o' rosies from being the subject of a million eager "hey, did you know that . . ." conversations.

Urban legends are not the only tales that carry on in the face of facts. In 2009, a study released in *Sociological Inquiry* looked into beliefs that Saddam Hussein had been involved with the attacks that occurred in New York and Washington, D.C., on September 11, 2001. In the months leading up to the invasion of Iraq, the Bush administration intimated such a relationship on many occasions, but as the war carried on and the spotty evidence behind any connection disappeared on close examination, members of the administration issued increasingly clear statements that Saddam had not been involved in planning or providing for the terrorist attack. However, when researchers talked to those who had initially believed that Saddam was involved, the later denials did little to shake their faith. They held onto their belief in the face of any amount of contrary evidence. Even when these believers were shown numerous statements from authoritative sources indicating that no such connection existed, the false ideas could not be dislodged. The unshakable quality of their belief was particularly evident in those who considered the connection between Saddam and 9/11 to be the justification for the American invasion of Iraq. Surrendering the connection would be surrendering their belief in the validity of the war, and so . . . they didn't. They wouldn't give up their belief in a connection even when shown video of President Bush personally stating there was no connection.

Surrendering a belief is always hard, especially if it fits with other preconceived notions. Whether it's a study that proclaims [Your Favorite Food] is the secret to a longer, happier life or a pundit who rains fire on politicians you don't happen to like, it's always enjoyable to have your private beliefs "confirmed."

But the tendency to hold onto beliefs *no matter what* can

have enormous negative consequences. For example, over the last decade, the evidence that human-generated pollution is significantly altering Earth's climate has grown increasingly convincing. Far more evidence exists for this change than for many commonly held beliefs. Yet climate change still meets with incredible resistance.

Why? Partly the opposition to global warming is due to the wealthy, powerful forces involved in seeing that our understanding of climate change remains muddied. The expenditures of organizations devoted to confusing the issue *far* exceed all the funds that go into researching climate change. Besides, accepting the idea of climate change means accepting the idea that what we do as individuals negatively affects the lives of billions around the world. That's an uncomfortable idea, especially if you really, *really* want to buy that new SUV. So climate-change deniers clutch stubbornly at the smallest straws and make illogical leaps that they would never make if they weren't so emotionally invested in the issue. Giving up their long-held beliefs would require admitting that they've been wrong. So they don't.

The same kind of opposition also holds true for ideas about evolution. The generally accepted notion is that everyone believed in a literal interpretation of the Biblical story of Genesis until Darwin came along and stole the imagination of scientists and a portion of the less-religious public. Very little of that idea is true. We think that there was a visceral reaction by religious groups against Darwin when his theory was first published because we see that kind of response in some quarters today.

On the Origin of Species gave birth to a popular story of an unpleasant debate between one of Darwin's supporters, Thomas Huxley, and Bishop Samuel Wilberforce. By most accounts today, Wilberforce was disdainful of the whole idea of evolution, unacquainted with Darwin's ideas, and sneeringly asked Huxley whether he was descended from apes on the side of his grandmother or his grandfather. Huxley then replied that he would sooner be descended from an ape than from a man who used his gifts to hide the truth. However, that version of the debate does not match with accounts written at the time. The truth is that "Soapy Sam" Wilberforce, one of England's most popular speakers, was not explicitly opposed to the idea of evolution. He published a critique of *Origin* on a scientific basis (a critique that Darwin found particularly clever), and the bishop's encounter with Huxley (which included other supporters of Darwin and was more of a roundtable discussion than a one-on-one debate) ended with all the men going out together to share a friendly dinner.

The Wilberforce-Huxley debate could serve as a proxy for the myth of how Darwin's ideas were received. The relationship between science and religion was in a state of flux throughout the nineteenth century. The broad acceptance of creation stories wasn't a mark of steadfast belief, it showed only a lack of alternatives. Just as Charles Lyell had been able to separate his proposals on the age of Earth from his beliefs as a Christian, most Christians at the time of *Origin* did not immediately view Darwin's work as a great threat to religion. The kind of literalism we live with today is an outgrowth of the type of fundamentalist Christianity that took hold only in the decades after Darwin's work was published. More importantly, ideas about evolution were in public circulation well before Darwin published any of his books. The picture of Victorian ladies fanning themselves to ward off fainting after being subjected to a glimpse of Darwin's theory is as nonsensical as the plague story behind ring o' rosies.

Theories of evolution—including human evolution—were not

only widely covered, but popular well before Darwin mentioned natural selection. So long as they were the right kind of theories.

The greatest champion of evolution in the decade before *On the Origin of Species* was not a scientist, but a young editor for *The Economist* named Herbert Spencer. Spencer was a member of a family that seemed to be chockablock with radical ideas, commercial success, and brilliance. He dabbled in half a dozen fields from engineering to medicine before taking up his pen as an advocate of free trade. In his 1851 book, *Social Statics,* Spencer wove together laissez-faire economics, sociology, and his version of Lamarckian evolution. Spencer's evolution was a force that drove toward improvement and complexity, a force that rewarded the "best" and punished second raters. It was in this book that Spencer coined the term *fitness* when talking about applying evolution to society. In his view, evolution was not only improving the human species generation by generation but also adapting it to live in a modern mercantile society. Only one stumbling block slowed the continuing improvement—government.

Spencer maintained that government was a primitive feature of human life—something that man had needed before the benefits of the market economy were clear. Since free trade economics was altering the structure of society, government was no longer needed. In fact, what government still remained was acting as an obstacle to human progress. The only thing that could be done was to harness government to the task of repairing the damage, even while that government was whittled away and discarded. Government's last task would be to oversee its own destruction. Ayn Rand, Grover Norquist, and Ron Paul would have loved him. And in fact, Spencer still has a strong following among certain stripes of free-marketeers who like to read his early, down-with-the-state positions.

Social Statics was a runaway success on both sides of the

Atlantic. Spencer's work was read and debated in backrooms and in newspapers. He was elevated to one of the most important figures in English philosophy. In 1855, Spencer published his second book, *Principles of Psychology*. In this work he argued that memories and ideas were physical features of the human mind, and that these features could be passed along to offspring through Lamarckian mechanisms. So if you were smart, it was because your parents were themselves learned, and you could count on your own children to inherit the expanded brain your hard work would deliver to them. *Principles* was not the best seller his first book had been, but Spencer remained a popular lecturer and writer, and the ideas of the book spread widely.

Swerving from economics to psychology may seem a large shift, but Spencer was seeking a "universal law," a sort of Grand Unified Theory of society, to describe the behavior of everything from atoms to markets. Over the next six years, he developed a system and presented an overview in his work *Progress: Its Law and Cause,* which was published in 1857, two years before *On the Origin of Species*. In this work Spencer elaborated on his theory of evolution, arguing that not just life but all structures in the universe had come from some undifferentiated, formless origin (not quite a Big Bang, more of a Big Soup). This idea wasn't unique to Spencer. The idea that everything from morality to nebulae all existed under the same set of natural laws was common in both science and society, and many people were looking for its underlying mechanisms. In searching for this key to the universe, Spencer expanded Lamarck's idea of evolution into a role that was almost godlike—driving everything toward, if not perfection, at least infinite, unending improvement. When it came to how this rule worked in biological systems, Spencer embraced Lamarck's use-it-to-improve-it concept.

By the time Darwin's work on natural selection appeared

at the end of 1859, Herbert Spencer was the most popular philosopher in England, but he was a philosopher with a problem. Although Spencer's reworking of Lamarckian evolution and universal development might make for a good read, and although it fit well with preconceived notions of how the world *ought* to work, it was long on philosophy, short on evidence. On the other hand, Darwin had spent 30 years refining his ideas—thinking of arguments and counterarguments and expanding on what his theory had to say about development, distribution, the changes seen in domesticated animals, the effects of climate, relationships between predator and prey, and much more. Darwin was one of those rare scientists who was able to put himself in the place of those who scoffed at his ideas, to argue against his own theory, and to provide a solid answer to those criticisms. He predicted many of the arguments skeptics would mount and tore them down step by step. There were then, and are now, those who doubt evolution, but for those who accepted the idea of evolution based on evidence found in the natural world, Darwin's theory was clearly superior to previous efforts. His book was an immediate hit. Almost half the first print run of *Origin* was snatched up by "circulating libraries" (a sort of Victorian Netflix) that mailed the book from one reader to the next. Though the first print run of Darwin's book was barely over 1,000 copies, it was likely read by 20 or 30 times that many people.

Spencer soon saw that his own thoughts on Lamarckian evolution were taking a beating in the marketplace of ideas. To hold off criticism he would have to absorb Darwin's natural selection into his own theories, but he didn't like it. Not at all. His version of evolution was purpose-driven and rewarding of effort. Darwin's theory depended on change that was merely . . . change, not a prize for past achievement. Worse, Darwin's theory didn't allow for improvements acquired in your lifetime to be passed along, so no matter how many companies you founded or how many millions you accumulated, you couldn't slide your Midas touch along to your offspring. Worst of all, natural selection threatened to upend everything Spencer had written.

Because Spencer is often cited as one of the founders of Social Darwinism, his ideas on evolution are often assumed to have been closely aligned with Darwin's. But Spencer's solution to how evolution worked radically opposed Darwin's explanation. Spencer believed that the mechanism behind evolution drove the system toward continuous improvement, with the ultimate goal of a system "in equilibrium" with his market-friendly, government-free people at the top of the system. In natural selection as presented by Darwin, creatures did change as they adapted to differing environmental conditions, and complexity might arise from this change. But increased complexity wasn't required. Neither was "equilibrium." The change was not required to be in the direction of "general improvement"—in fact, the idea of general improvement didn't exist. Those animals that survived to have offspring might well have been smaller and weaker, if that was what the environment demanded. And Darwin had the facts on his side.

If the universal rule that Spencer was still expounding was going to survive the introduction of natural selection, it would have to do so by swallowing up Darwin's theory. In a neat example of literary adaptation, Spencer took Darwin's terminology, made it his own, and then proceeded to ignore the actual ideas at the heart of Darwin's work. Spencer's editing of Darwin turned evolution into a mishmash of Lamarckian mechanisms and natural selection catch phrases. Accuracy gave way to popularity and the perpetuation of classism and racism.

In 1865, Spencer published his next work, which this time

included his altered version of Darwin. *The Principles of Biology*, written by a laissez-faire economist who believed that human evolution would lead to an increasing separation of the species into "divisions of labor," became one of the founding documents of Social Darwinism. Most of the images and ideas that persist about how evolution shapes human society didn't come from Darwin, they came from this work by Spencer. Even the phrase most associated with Darwin's theories—"the survival of the fittest"— wasn't coined by Darwin. That was Herbert Spencer.

But you can't shake an idea that clicks. Spencer's theory was much more palatable for those who were used to the Great Chain of Being. It retained the order, the drive, and the neat location of man on top and apart from the "lower" animals. Darwin's views on natural selection proved right; Spencer's ideas on Lamarckian inheritance proved wrong. But you wouldn't know it from the influence each man had on society. The spread of Spencer's *Social Statics* was such that Justice Oliver Wendell Holmes cited Spencer's work in his dissent of a Supreme Court case on the Fourteenth Amendment.

Spencer's ideas stick because they fit the built-in prejudices and concerns that the well off have always had about immigrants and the poor. In his writing, he managed to invent Social Darwinism before there was such a thing as Darwinism, so it seems only appropriate that among Spencer's followers was *The Time Machine* author, H. G. Wells. Wells based his vision of future society on Spencer's work: the childlike Eloi and the trollish Morlocks were the end results of Spencer's division of labor driving humanity into separate species.

Unfortunately, the distorted idea of evolution that Spencer created didn't vanish from history—it shaped history. This wouldn't have been such an issue if it had been known as "Social

Spencerism," but because Spencer's ideas and others like them were bundled under the Social Darwinism label, they helped shape public perception of natural selection into a theory Darwin wouldn't recognize. This idea, like an urban legend, never seems to die.

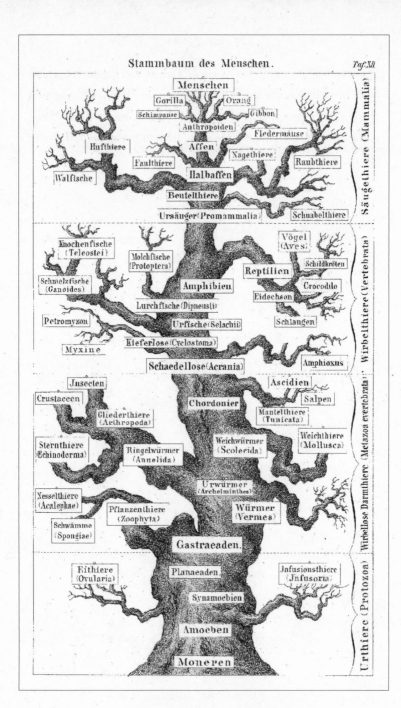

Stammbaum des Menschen.

Taf. XII

Menschen

Gorilla · Orang
Schimpanse · Gibbon
Anthropoiden
Affen · Fledermäuse
Hufthiere · Nagethiere
Faulthiere · Raubthiere
Walfische · Halbaffen
Beutelthiere
Ursäuger (Promammalia) · Schnabelthiere

Säugethiere (Mammalia)

Knochenfische (Teleostei)
Molchfische (Protoptera)
Vögel (Aves)
Schmelzfische (Ganoides)
Schildkröten
Amphibien
Reptilien
Crocodile
Lurchfische (Dipneusti)
Eidechsen
Petromyzon
Urfische (Selachii)
Schlangen
Myxine
Kieferlose (Cyclostoma)
Schaedellose (Acrania)
Amphioxus

Wirbelthiere (Vertebrata)

Jnsecten
Ascidien
Crustaceen
Chordonier
Salpen
Gliederthiere (Arthropoda)
Mantelthiere (Tunicata)
Sternthiere (Echinoderma)
Ringelwürmer (Annelida)
Weichwürmer (Scolecida)
Weichthiere (Mollusca)
Urwürmer (Archelminthes)
Nesselthiere (Acalephae)
Pflanzenthiere (Zoophyta)
Würmer (Vermes)
Schwämme (Spongiae)
Gastraeaden.

Wirbellose Darmthiere (Metazoa evertebrata)

Eithiere (Ovularia)
Planaeaden.
Jnfusionsthiere (Jnfusoria)
Synamoebien
Amoeben
Moneren

Urthiere (Protozon)

The Pedigree of Man

ERNST HAECKEL, 1874

CHAPTER 7

Flim-Flam Men

> Darwin's original contention, that biological evolution is a
> natural process, affected primarily by natural selection, has
> thus become increasingly confirmed and all other theories
> of evolution requiring a supernatural or vitalistic force
> of mechanism . . . together with all Lamarckian theories
> involving the inheritance of acquired characteristics, have
> become increasingly untenable.
>
> JULIAN HUXLEY, *Evolution: The Modern Synthesis* (1942)

As the owner of a patent medicine company, Louisiana state
congressman Dudley LeBlanc might have been expected to take
some of his own "pain powder" when he injured his big toe.
Instead he paid a visit to the doctor, where he was given a mixture that included a dose of B vitamins. LeBlanc swiped a bottle
of the doctor's concoction, ran off, and used it as the basis of his
next big idea.

LeBlanc was badly in need of a fresh start. The 1940s was a
tough time for small-time tonic peddlers. The medicines offered

by pharmacists were increasingly more effective than products sold off the back of a truck, and competition from large manufacturers was making it harder and harder for local producers to sell even common medicines such as cough syrup. And there was a bigger problem. After decades of almost anything goes, many of the ingredients that had made patent medicines popular in the past—such as tinctures of opium and cocaine—were likely to draw the ire of a more and more vigilant U.S. Food and Drug Administration (FDA).

That increase in government watchfulness was the problem plaguing LeBlanc. The products of his Happy Day Company (including Dixie Dew Cough Syrup and Happy Day Headache Powders) had been seized for being ineffective and for containing dangerous ingredients. He needed a new product. Based on the doctor's pain remedy and some creative thinking, LeBlanc produced his next creation—Hadacol.

Hadacol was a brown, syrupy patent medicine with a rather vile taste and an ingredient list that included dilute hydrochloric acid. It was sold at a high price and, even though it was advertised for many purposes, wasn't particularly effective against any illness. Still, bottles and flasks of Hadacol sold by the millions.

In fact, Hadacol would become something of a national fad, complete with a Hadacol boogie, Hadacol cocktails, and a traveling Hadacol Caravan that featured a full-sized circus along with stars such as Bob Hope, Hank Williams, and Judy Garland. Two factors made this possible: one inside the bottle and one outside.

First, LeBlanc was a matchless salesman who primed the advertising pump with every dime brought in by the company. He carefully skirted the FDA by using "testimonials" rather than direct claims, but those catching a Hadacol radio ad could hear stories of the goopy fluid curing rheumatism, anemia, heart disease, stomach ulcers, and cancer. In addition to being promoted by a touring show full of clowns, acrobats, music, and stars, Hadacol became one of the most advertised products on radio. LeBlanc produced endless schemes and promotions—such as sending coupons good for a free bottle into areas where pharmacies didn't yet carry the mixture—to drum up demand.

The other big secret of Hadacol was the "preservative" that was "used to keep the vitamins potent." That preservative was alcohol—12 percent alcohol. At a time when much of the southern United States was officially "dry," Hadacol offered an alternative to a visit with the local bootlegger or a long trip into the city. The dash of hydrochloric acid in the Hadacol formula dilated arteries, allowing the alcohol to enter the blood stream more quickly and giving the concoction a bigger "kick" than might otherwise be expected from the 24-proof contents. By 1951, despite having no medicinal properties, and despite being railed against by the American Medical Association, Hadacol was one of the best-selling "medicines" in the United States.

Good packaging and promotion can work not just for products, but also for animals. One obstacle to Darwin's ideas on natural selection was the presence of so many creatures with adaptations that seemed so profoundly *unhelpful*. The tail of a peacock doesn't help the bird run faster, or find more food, or fly farther, for example. In fact, just the opposite is true. Carrying around such a collection of feathers (which actually aren't the tail feathers at all but are greatly elongated forms of the feathers located above the tail quills) must make the peacock clumsier when hunting for the small animals it eats. Why does such an elaborate burden make the peacock more "fit"? Because peahens like them.

Peacocks are just one of hundreds of thousands of species where coloration, behavior, and even basic form have been strongly shaped by the struggle to gain a mate. After all, an animal that's fit for its environment in every other way but that fails to pass along its genes to the next generation has no effect on shaping the future of its species. The result is that fitness in any given situation isn't just securing food and warding off disease; it's the ability to gather a mate (or mates) and produce healthy offspring. Natural selection is often seen as a fight that takes place on Tennyson's battlefield of "nature red in tooth and claw," but it's also a struggle that happens in the nest, the burrow, and the bedroom.

The effect of this struggle can be seen in nearly every type of animal where there is sexual reproduction. Many creatures carry flashy displays intended to catch the eye of potential mates. These may serve as a kind of secondary display of fitness—Hey, baby, if I can survive in the jungle with a tail *this big*, just think how fit I must be!—but that's not their direct function. A sexual display is about securing sex. If that sexual display gets so burdensome that it starts to interfere with other essential aspects of survival, then the nonsexual aspects of natural selection will take care of that issue. Likewise, if a female (or male) fixates on features in a prospective mate that impose a huge penalty on the offspring, another branch of life's shrub is likely to meet with God's own garden shears.

Still, sexual selection can lead to some spectacular developments. One feature that has occurred again and again is the elaborate headgear males use in an effort to impress females. Many animals, like stag beetles or stags of the deer variety, use these head ornaments in battles that can range from ritual taps to deadly contests of strength and persistence. Others, like hornbill birds, use them strictly as visual displays.

Many dinosaurs also carried fancy headgear. One of the most familiar species of dinosaur is the three-horned *Triceratops horridus*. The bony neck frill and massive set of horns paired with the large heavy body of this herbivore have from the outset put people in mind of a rhinoceros. Many people have assumed that Triceratops also behaved like a rhino, using its horns to ward off attacks from large predators. It seems logical, considering the size and orientation of the horns. However, Triceratops is only one of a group of dinosaurs that carried such "weaponry."

Many Triceratops relatives had one horn, two horns, or a dozen horns that could be oriented forward, sideways, or backward; they could even be shaped like hooks. The neck frills on some of these dinosaurs were delicate, certainly too delicate to have acted as shields against a hungry *Tyrannosaurus rex*. More study has indicated that the Triceratops head structures were probably not used for defense but related to sexual display. One big clue is that, just as deer go from simple to branching horns as they mature, ceratopsian dinosaurs had headgear that changed drastically as the animals aged (so much so that what are now recognized as juvenile examples of some dinosaurs were previously mistaken for completely different species). Such headgear may have been used in the kind of tussles that many mammals go through to establish their dominance, but even that appears to be unlikely. Just as the head structures were too fragile to act in defense, they were also too fragile to support more than ceremonial jousting. More than likely, they were used just like the display structures on the dinosaur's relatives, the birds. The structures existed to impress potential mates.

Humans are also susceptible to sexual displays that may have nothing to do with being fit in any other way—hairstyles, clothing, pom-poms, and Porsches. And sex isn't the end of it; we give in to signals of all sorts. Plants long ago developed fruit with bright colors and sweet aromas as an enticement for hungry primates. Those same colors are appealing on bananas and on magazine covers. Advertisers of all sorts have learned that you can sell any product with the right pitch, even if the product is purely craptacular. It worked for Hadacol, it works for diet pills today, and it worked for a man named Ernst Haeckel.

What Haeckel was selling was his own brand of evolution. He was a Prussian professor of anatomy who originally specialized in worms, sponges, and other soft-bodied organisms. He actually came to visit Darwin at his home in 1866, exchanged correspondence with him over a period of years, and professed to be a follower of natural selection. But if Haeckel studied natural selection, he ignored much of what he read when it came time to write down his own ideas. Like Lamarck, Haeckel restructured the classification of animals (giving us the terms *phylum, phylogeny,* and *protista*). Haeckel also developed a theory of evolution that was based on the retention of acquired traits along with a drive toward increasing complexity.

On the merits of his arguments, Haeckel's ideas were no more than a variation on the version of evolution being pushed by Spencer and many others—in fact, they were not much more than an update of what Lamarck had suggested a century earlier. Even if this kind of Great-Chain evolution was popular with the public, naturalists might have been expected to scoff at an idea based on theories that had long proved unsatisfying. But Darwin's version of evolution was far from the only one that enjoyed academic support.

The lack of a clear, demonstrated mechanism for the inheritance of traits left natural selection as just one of several theories with adherents in scientific circles. Darwin himself had suggested that inheritance occurred through small packets of information carried in the blood. This suggestion represents one of the few times that Darwin speculated with little experimental basis. Darwin was wrong, and experiments devised to test his theory showed he was wrong. Evolution itself was not in doubt—extinction, change, and the appearance of new species was clear—but even those who supported Darwin's theory of natural selection had begun to doubt that it was the main mechanism behind evolution. As always, it seemed too simple. The neo-Lamarckians were in ascendance, and Ernst Haeckel was about to become a star.

First, Haeckel would make a discovery that established evolution from a direction no one else had expected. In medical school, Haeckel had been fascinated by embryology—the study of development of organisms before their birth. Returning to that fascination, he looked at the developing embryos of several species and made an astounding discovery: in the process of development, each species went through a sort of replay of its evolutionary history. As it developed, the embryo of a mammal would spend some time as a sphere of undifferentiated cells—very like an early colonial organism. As the embryo became more organized, it would become more fishlike and then pass through stages that were amphibian and then reptilian before showing definitive signs of its mammalian heritage. Even finer levels of differentiation could be detected. Haeckel published a set of carefully rendered images tracking the development of embryonic fish, chickens, dogs, turtles, salamanders, cows, and humans. At their early stages, these embryos were all very similar, but the fish became more differentiated in the fish-like stage, the turtle halted its development and became more specialized in the reptile stage. Only after that point did the chicken become different from the mammals. The dog, cow, and human embryos continued to be very similar until they reached the point of mammalian development at which the types of creatures split apart.

Haeckel had discovered that the developing embryo acted like a biological DVR. Careful study of embryonic development was like holding down the fast-forward button on millions of years of evolution. Haeckel's book and sketches of the embryos caused a sensation. His term for this encoding of evolution into developmental stages was translated as "ontology recapitulates phylogeny" (development demonstrates the nature of relationships)—a phrase that would trip from the tongues of biologists for decades to come. Haeckel had opened a new window for studying evolution. Naturalists were no longer forced to string together a patchwork history of fossil forms to determine the ancestry of a creature. Recapitulation showed that answers were encoded in the creature itself. Darwin was thunderstruck by this discovery and wrote that, had Haeckel written his work first, there would have been no need to write *On the Origin of Species*. (However, Darwin's book predated Haeckel's and did not, as some of Darwin's critics have claimed, make use of Haeckel's data.)

This was just the first of Haeckel's grand achievements. In his 1871 book, *The Descent of Man*, Charles Darwin had suggested that all humans came from Africa, and that they shared a recent common ancestor. Darwin's reasoning was straightforward. Anatomical comparison showed that humanity's relatives were the Great Apes. In particular, the gorilla and the chimpanzee seemed to be our closest living kin. Gorillas and chimpanzees lived in Africa, and only in Africa. If humanity had shared a common ancestor with these primates, that ancestor likely also lived in Africa. Further, despite superficial differences, human beings didn't demonstrate a high degree of variation. Compared to many other animals, individual humans were remarkably similar.

Darwin's book classified all humans as a single species and considered the differences of race to be minor. In some ways, Darwin was as bigoted as any other man of his time. He didn't doubt for a moment that the men on board the *Beagle* were infinitely superior to the natives they met in Tierra del Fuego when it came to their *cultural* accomplishments. But he did not believe that one race was physically superior to another. Darwin felt that the capabilities of all humans—physical and mental—were probably very similar. Those who migrated from Africa long ago differed because they had adapted to local conditions.

> Although the existing races of man differ in many respects, as in colour, hair, shape of skull, proportions of the body, etc, yet if their whole organisation be taken into consideration they are found to resemble each other closely in a multitude of points. Many of these points are of so unimportant or of so singular a nature, that it is extremely improbable that they should have been independently acquired by aboriginally distinct species or races. The same remark holds good with equal or greater force with respect to the numerous points of mental similarity between the most distinct races of man. The American aborigines, Negroes and Europeans differ as much from each other in mind as any three races that can be named; yet I was incessantly struck, whilst living with the Fuegians on board the *Beagle*, with the many little traits of character, shewing how similar their minds were to ours; and so it was with a full-blooded negro with whom I happened once to be intimate.[1]

Haeckel's view of humanity could not have been more dissimilar. He felt that the differences among the races were more than enough to break human beings into separate species. In fact he divided humans, even when they were physically very similar, along lines of language groups. Haeckel not only didn't believe all men belonged to the same species, he didn't believe they *came* from the same species. He theorized that each group of modern humans had evolved separately from speechless *Urmenschen*, and, as separate products of evolution, each race represented a species with a different "potential" that was encoded into its language as well as its biology. Of course, he had ideas about which races held the most potential, and he didn't hesitate to put these "species" in order.

> The Caucasian, or Mediterranean man (*Homo Mediterraneus*), has from time immemorial been placed at the head of all the races of men, as the most highly developed and perfect. . . . This species alone (with the exception of the Mongolian) has had an actual history; it alone has attained to that degree of civilization which seems to raise men above the rest of nature.[2]

For Haeckel, only Europeans—and the Aryans he believed to be the ancestral "species" that had given rise to Europeans—were real people. Everyone else on Earth was not just inferior, they were *animals*. Herbert Spencer's Social Darwinism had created a version of evolution that was designed to preserve the social hierarchy and promote laissez-faire economics as a biological imperative. Haeckel's version of human ancestry went further. It neatly enfolded German romanticism and the myth of the Aryan race.

Again, you might think that scientists would be disinclined to support a theory that was so laden with racism, classism, and convenient confirmation of what the elite so desperately wanted to believe. And they might have, but Haeckel was about to get spectacularly lucky.

In predicting that the common ancestor of all human beings had come from Africa, Darwin suggested that those looking to find fossils of human ancestors should look for them in Africa. But at that time there were no fossil ancestors. All of Darwin's theory was based on a series of logical conclusions, but they were theoretical. There were no bones to back up his story.

Haeckel, following the mythology of the Aryan race, pointed to another site. The ancestors of *Homo Mediterraneus* should be sought in Asia. Where Darwin stopped at the idea that common ancestors had existed and made only few conjectures about their nature, Haeckel's speculation came with a precise number. There would be, he predicted, exactly 22 steps between man and apes. In the midst would be a creature that was nonverbal, and physically halfway between apes and man. When this halfway point was found, the creature would have a smaller cranium than a modern human but features that separated it clearly from other apes. It would be "the missing link"—a phrase that Darwin never used, and that extended back to the Great Chain of Being with its ordered ranks of creation. Even though this creature was entirely the creation of Haeckel's imagination, he gave it a formal name: *Pithecanthropus* (ape man).

In 1891, Dutch anatomist Eugène Dubois, following Haeckel's advice, went looking for the ape man in East Java. Along the banks of the Solo River, he found a portion of a skull that was smaller than that of a modern man, but quite a bit larger than any ape's. Along with it he found a femur that also differed from that of modern humans. It was clearly an earlier form of man, and Dubois dutifully named it *Pithecanthropus*. Haeckel was triumphant, and Darwin had been dead for a decade.

There was one final factor that elevated Haeckel with both the public and his fellow scientists—he could draw beautifully. Haeckel's book *The History of Creation* was first published in 1868; new versions came out regularly for another three decades. The book begins with a brief nod to Darwin, before pulling back the camera to show that what Darwin suggested was only a small piece of what Haeckel was assembling. Like Herbert Spencer, Haeckel sought a unified theory of biology and society.

> The intellectual movement to which the impulse was given
> thirty years ago by the English naturalist, Charles Darwin, in
> his celebrated work *On the Origin of Species*, has within this
> short period assumed dimensions of unprecedented depth and
> breadth. It is true that the scientific theory set forth in that work,
> which is commonly called Darwinism, is only a small part of
> a far more comprehensive doctrine—a part of the Universal
> Theory of Development which embraces in its vast range the
> whole range of human knowledge.[3]

The grand flowery language was accompanied by dozens of wonderfully sketched plates, each of them executed in a style that infused even jellyfish and oceanic plankton with almost heavenly perfection (see Figure 7.1). Haeckel's book outsold Darwin's work year after year. His book was distributed in the United States up until the 1930s (the point at which championing people who droned on about the superiority of the German race became a bit less popular).

Haeckel's position was slow to crumble. As early as 1874, some of those who worked with Haeckel pointed out that the embryonic drawings were a little *too* perfect. The similarities between embryos were exaggerated, and the stages through which embryos passed was not nearly so clearly defined as Haeckel had

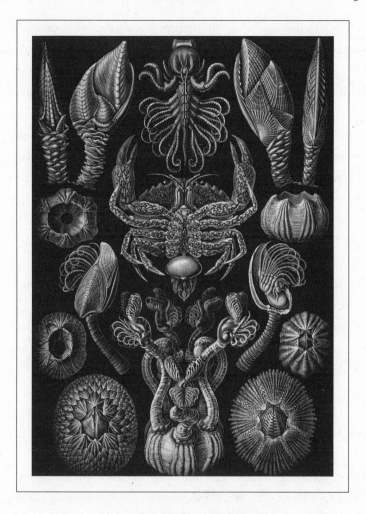

FIGURE 7.1. CIRRIPEDIA

ERNST HAECKEL, 1904

suggested. Haeckel had used more than a little artistic license to "discover" this high-speed review of evolution. Recapitulation-ism would hold on for decades as science, and even longer as popular knowledge, but by the early part of the twentieth cen-tury, scientists would realize that embryos did *not* retrace a species' path through time. Similarities among the development of embryos were an inherited feature. Just as organs that performed the same function in creatures as different as barnacles and men, embryonic stages performed their role in preparing a developing organism for birth.

The Asian origin of modern man would take even longer to fall into disrepute. As scientists uncovered more specimens, they finally realized that what Dubois had found (now known to be partial remains of *Homo erectus*) was not a species that had evolved in Asia, but one that had moved there. Far from being the home of early man, Java had marked the extent of his range. Earlier fossils of *H. erectus* were found in Africa, and it became clear that the species had originated in Africa and spread to parts of Europe and Asia hundreds of thousands of years before the emergence of modern man. The earliest specimens of modern *Homo sapiens* were also found in Africa, indicating that Darwin had been correct in his prediction of a recent, common, African ancestor for all humans.

Despite the fossil evidence, the idea of some humans evolving in Asia rather than Africa held on decades longer as part of the "multiple origins" idea, the theory that humans had evolved in more than one location and had more than one ancestor. Some anthropologists thought they detected distinct features among early *H. sapiens* fossils from different regions that indicated that our species had come from not one ancestral species, but several (a pattern unknown in other creatures). This Haeckel-inspired theory had vocal adherents right up until the 1990s, when genetic studies showed that the out-of-Africa theory—Darwin's theory— was right all along. Humans shared one common evolutionary history, and until quite recently that history had taken place entirely in Africa.

Haeckel's recapitulation theory had been based, at best, on a mistake and at worst, on fraud. His suggestion that humans had originated in Asia was based on ideas that went back into a mythic version of German pre-history. With the initial finds in Java, he got very lucky (it didn't hurt that the only person seriously looking for hominid fossils was looking where Haeckel sent him).

Some recent biographies have made an effort to redeem Haeckel's reputation from the connections made between his ideas and everything from eugenics to Nazism. Certainly Haeckel was a prodigious author and produced many worthwhile results that included a better understanding of how species cooperate and compete. (The word *ecology* was another of Haeckel's coinages.) His illustrations are undeniably luminous, and Haeckel was dead well before fascism swept to power in Germany. But there is no doubt that Haeckel's ideas contributed directly to the use of science as a tool to forward racism. Haeckel had a substantial role in creating the Monist League in 1904, an organization dedicated to maintaining "racial hygiene." For all his genius, for all his genuine discoveries, Haeckel was guilty of going far beyond the data to introduce flights of fancy like his fixed set of 22 steps between man and ape. Worse, he used these pseudoscientific inventions to lend credence to some of humanity's worst impulses at a time of rapidly increasing tensions.

Not only did the apparent triumphs of Haeckel's work end up being grand misdirections, it's no accident that the period of his greatest popularity coincided with a marked increase in friction between evolutionary biology and religion. Even though Haeckel's own ideas contain a strong thread of mysticism and he eagerly incorporated the mythology of Aryanism, Haeckel detested all forms of religion. He rarely missed an opportunity to portray Christians as the worst sort of simpletons or Jews as subhuman. His assault on organized religion was often crude, spiteful, and filled with as much prejudice as he showed about race. Haeckel might not have been the first to throw a punch in the fight between religion and supporters of evolution, but he threw them enthusiastically and often.

By the 1930s, not only had Haeckel fallen out of favor with the public, but Darwin's ideas were enjoying a great resurgence with scientists. Mendel's lost notebooks on inheritance had been rediscovered at the start of the twentieth century, and genetics was filling in the gaps on how features were passed from one generation to the next. Studies on population and ecology were examining the roles of both individuals and groups in the competition within and among species. The Lamarckian style of evolution favored by Haeckel, Spencer, and the Social Darwinists did not fit the results. The flood of new evidence supported natural selection as the primary driver of evolution.

In 1942, Julian Huxley (grandson of Darwin's friend and supporter, Thomas Huxley) published the book *Evolution: The Modern Synthesis*. In it he laid out a series of simple rules that united genetics and population dynamics with natural selection. More than 60 years later, the "modern synthesis" is as modern as ever. All efforts to propose alternative means of species origin (and there have been many) have fallen by the wayside. Genetic testing has shown that natural selection isn't just one factor among many driving evolution but is an *overwhelming* force that greatly exceeds all other causes combined. Some of Ernst Haeckel's ideas persist today as part of the vague public understanding of evolution, but in the end much of his work was smoke, mirrors, and some really nice drawings.

Hadacol also turned out to be even less than it seemed. When LeBlanc sold the company in 1951, the new managers were shocked to discover that the $3 million in sales had come at a price of $5 million in advertising. Despite what appeared to be a thriving business, the company had actually been losing money for some time. LeBlanc did not seem too disturbed. Asked by comedian Groucho Marx just what Hadacol was good for, LeBlanc gave a truthful answer: "It was good," he said, "for five and a half million for me last year."[4]

Eugenics Congress Announcement

Number 1. History and Purpose of the Congress.

EUGENICS IS THE SELF DIRECTION OF HUMAN EVOLUTION.

LIKE A TREE EUGENICS DRAWS ITS MATERIALS FROM MANY SOURCES AND ORGANIZES THEM INTO AN HARMONIOUS ENTITY.

Third International Eugenics Congress

New York City, August 21-23, 1932.

*Announcement of the Third
International Eugenics Congress, 1932*

CHAPTER 8

Big Mike and
the Paper Hanger

> We drown the weakling and the monstrosity. It is not
> passion, but reason, to separate the useless from the fit.
> SENECA (63)

Bananas have large, hard seeds. Lots of seeds, actually—at least
the undomesticated varieties still found in Southeast Asia are
loaded with hard seeds. When the plant was originally domesticated (probably over 7000 years ago in Papua New Guinea), they
were grown from those seeds. However, you're unlikely to be surprised to hear that commercial bananas today are lacking seeds.
Like seedless grapes, watermelons, and blackberries, they are
reproduced through some other means. In the case of bananas,
commercial varieties are propagated from cuttings taken from a
parent plant that was a demonstrated producer of large quantities
of desirable fruit. So most bananas sold in the world come from
"cultivars" of a single plant—clones of a single seedless parent.

If you were born before the 1950s, you may have eaten two

bananas in your life. If you're younger, chances are you've eaten only one. Of course, you may be fond of plantains, or be lucky enough to live in an area where your grocery provides more variety, but for most of America and Europe *banana* actually means *Vietnamese Cavendish banana*. That fruit is the large, yellow banana sold in these markets today.

Before the Cavendish, the bananas sold in most markets went by the name of *Gros Michel*—the Big Mike. Big Mike was carefully selected for a whole set of properties that made it a terrific banana for the market. It had a bright, attractive, relatively unblemished skin that wasn't just pretty, it was also tough. The yellow skin peeled easily but was thicker than the skin on other bananas, and Big Mike lasted for a longer period without becoming overripe. These qualities made it easier to ship the bananas without refrigeration. Many people who ate them in their childhood still have the memory of bananas being better back then—and they are right. It's not just nostalgia. The Gros Michel had a smooth texture, along with a taste that was sweeter and less starchy than today's Cavendish.

Breeding for market appeal isn't new and it isn't unique to bananas. Plants and animals of all sorts have been domesticated over the last 12,000 years or so. Though animal breeding may seem like something more likely to concern country farmers than urban professors, until quite recently, domesticated plants and animals played a very direct role in the lives of most everyone, even those who lived far away from farms. Not only were people familiar with animals as pets or as transportation, but until refrigeration and high-speed transportation made it possible to raise livestock far away, stockyards and slaughterhouses were common features of even the largest urban areas.

When Charles Darwin and Alfred Russel Wallace needed examples for the new theory of evolution through natural selection, they turned to domesticated animals because they knew their audi-

ence would find them familiar. Not only were domesticated animals everywhere, but anyone who took time to note the herds that
passed in the street could see that they were clearly not "fixed."
Each of the various animals used as food, labor, or pets came in
a wide array of varieties, each with its own special properties. A
few generations of selection by farmers, sometimes even just one
fortunate cross, could generate a new "breed" that had unique and
valuable characteristics. All the Angus cattle today can trace their
ancestry to a pair of hornless cows on a Scottish farm in the first
few decades of the nineteenth century. A wavy-coated retriever
crossed with the now-extinct Tweed Water Spaniel formed the
lineage for the millions of Golden Retrievers to follow. In the early
part of the 1800s, American farmers in New England kept flocks
of "otter sheep" with long bodies and legs too short to allow them
to jump the area's low stone fences. Not only did otter sheep come
into existence from breeding the offspring of a single animal that
had the short-legs and long-body form, the whole breed disappeared almost as swiftly only a few decades later when other breeds
became more desirable. Domesticated animals demonstrated that
the basic design of a creature could be altered, sometimes radically, within a few generations. These new forms could be wildly
successful or meet with extinction. So it was no wonder both Darwin and Wallace found them valuable examples. However, those
examples got some people thinking in another direction.

The first printing of Darwin's *On the Origin of Species by Means
of Natural Selection* was sold out before it reached the shelves—an
event all authors would love to experience at least once. Among
those first readers were several members of Darwin's family,
including a half-cousin, Francis Galton. Thirteen years younger
than Darwin, Galton had been a child prodigy who learned to
read at an age when most children were learning to speak and who
could converse in Greek and Latin by the time he turned five. But

for all his inordinate brain power, Galton had difficulty in school.
He had trouble taking direction and was prone to breakdowns that
plagued him through his life. Like Darwin, Galton made a run at
medical school before dropping out to take on another topic (in
Galton's case, mathematics). Unlike Darwin, Galton went back to
medicine, and then left college to pursue his own racing thoughts.
Like his half-cousin, Galton also spent some time tracking through
areas less trampled down by European civilization, and his book
Narrative of an Explorer in Tropical South Africa earned him some
of the same praise and attention Darwin had received when he
published his popular volume on the *Beagle* journey.

Some of that praise came from Darwin. The distance of their
relationship, and the difference in their ages, meant that Darwin
and Galton to that point had not been very close—a detachment
that's easy to detect in Darwin's letter to Galton:

> You will probably be surprised, after the long intermission of
> our acquaintance, at receiving a note from me; but I last night
> finished your volume with such lively interest, that I cannot
> resist the temptation of expressing my admiration at your expe-
> dition, and at the capital account you have published of it.
>
> . . .
>
> If you are inclined at any time to send me a line, I should very
> much like to hear what your future plans are, and where you
> intend to settle.
>
> . . .
>
> I live at a village called Down near Farnborough in Kent, and
> employ myself in Zoology; but the objects of my study are very
> small fry, and to a man accustomed to rhinoceroses and lions,
> would appear infinitely insignificant.[1]

Galton's adventures didn't end on his return to England. He
made contributions across multiple fields that survive to our day,

among them the use of fingerprints for identification. A great deal of the science of statistics was created by Galton. He devised the standard deviation, regression toward the mean, correlation, and normal distribution. He went as far as producing a mechanical device that took the place of a "random number generator" to demonstrate his ideas about distribution. Galton was the first to theorize about the "wisdom of crowds" when he observed a contest to guess the weight of an ox and saw that, while individual guesses might be wildly off, the mean of all the guesses was astonishingly close to the real value.

Still, though it's always a bit dicey to make a diagnosis over a span of decades on the basis of only writing and remembrances, it's tempting to wonder if Galton suffered from some disorder that might have a convenient label today. Obsessive-compulsive disorder, perhaps, or some problem akin to Asperger's. His bouts of emotion seemed to baffle those around him, his obsession with exactly counting and calculating values was expressed in nearly every field he entered, and even his ability to fixate on an idea for years seemed to come with a price. Galton appeared to recognize this and worried that he might be driven mad:

> Men who leave their mark on the world are very often those
> who, being gifted and full of nervous power, are at the same time
> haunted and driven by a dominant idea, and are therefore within
> a measurable distance of insanity.[2]

Though he helped shape many items that we still live with today, from the moment *On the Origin of Species* was published, there was one idea that dominated Galton's thoughts. If Darwin had been taken with Galton's recollections of his travels in Africa, Galton was immediately and intensely struck with what he found in Darwin's book. So smitten was he that it would become the focal point for the rest of his life. The overall argument for natural selection wasn't what gripped Galton. He was spellbound by one of the early chapters—one that centered on the variations of domesticated animals. If Darwin had hoped the use of these familiar animals would make good examples, they certainly worked to convince Galton—but maybe not quite in the way that Darwin would have wanted.

Throughout *Origin,* Darwin had carefully avoided the direct application of natural selection to the evolution of humans (he would tackle that topic head on in *The Descent of Man, and Selection in Relation to Sex*). Galton felt no such compunction. However, it wasn't humanity's origins that concerned Galton, it was humanity's future. He eagerly wondered whether human traits were inherited as they were with other animals, and he set out to answer the question in the way that he had answered others—through statistics. Height, weight, and hat size would all come under Galton's examination, but it was already clear enough that human beings could inherit physical traits—otherwise you would not end up with children who so often resembled the parents. He was more concerned about whether the less visible traits were inherited.

The question didn't interest Darwin. He considered most people to have more or less the same level of intelligence. Darwin regarded accomplishment as a measure of opportunity and hard work. In other words: all nurture, no nature.

Galton did not agree. He set out to look at the children of people who were "eminent" to see if they showed signs of inheriting the abilities of their notable parents. This first study of human intelligence and accomplishment would face the same obstacles faced by every such study that followed after—the question of whether a child is innately smarter because he or she inherited the trait from bright parents, or because of surroundings that include a respect for learning and opportunities to both acquire and express knowledge. To control for the effects of environment when you're examining the situation in person is hard enough, but Galton's study was done much more remotely. To find out if humans could pass along their intelligence, Galton read books. He examined biographies and records concerning eminent men and their descendants. He was aware that this method of study wasn't perfect. He suggested studies of twins and of adopted children, anticipating the way such studies would be done for more than a century to follow. However, the lack of such studies didn't stop him from searching for intelligence in the pages of books. And he found what he was looking for.

Galton's research was pioneering in many ways, and out of his obsession came mathematical tools that benefited researchers in all fields. The value of his statistical methods was quickly realized and the carefulness of his approach impressed many, including Darwin. However, there was nothing remotely "double blind" about Galton's research. Despite his studious tabulation and examination of his numbers, his study suffered from the extreme prejudice by which the lives of the notable were far more likely to be recorded (and their deeds to be inflated). There was no way to verify the validity of the praise heaped on some of the historical figures in his work, and no way to compare their intelligence with that of less "accomplished" men and women in any measurable way. Galton's study was highly mathematical, but it wasn't scientific.

By 1869, Galton had published his book *Hereditary Genius*. In the book he pressed the idea that there should be more attention given to favorable traits in human breeding. After all, if the form and behavior of sheep, cows, goats, and all other domestic animals could so readily be affected by the introduction of some control over their breeding, shouldn't the same apply to human beings? Wouldn't control over human breeding be required to weed out undesirable traits and ensure the survival of favorable traits? If we domesticated cattle, shouldn't we also domesticate man?

Darwin considered the question and responded that he thought that Galton was probably right in his basic ideas. In fact, Darwin admitted that the program of breeding Galton suggested might be the *only* way that these favorable traits could be ensured. However, Darwin viewed the idea as impractical. Instead, he suggested that people be told the importance of inheritance, and left to make their own decisions:

> I should be inclined to trust more (and this is part of your plan) to disseminating and insisting on the importance of the all-important principle of inheritance.[3]

The breeding of people for specific traits wasn't the only issue Darwin had with his half cousin's book. Galton, drawing conclusions from his swarm of statistics, had produced numeric evidence that some races of men were inferior not only mentally, but also physically. Even that some races were inherently uglier than others. Darwin saw no proof of this.

Finally, Darwin was surprised to discover that Galton had taken from *On the Origin of Species* an idea that ran completely counter to the whole design of natural selection. The subtitle of *Origin* was *The Preservation of Favoured Races in the Struggle for Life*. Galton took this subtitle quite literally. Individuals, he declared, did not matter. It was the race that was valuable.

> The life of the individual is treated as of absolutely no importance, while the race is as everything; Nature being wholly careless of the former except as a contributor to the maintenance and evolution of the latter.[4]

His words on the subject seem particularly chilling in retrospect. The idea was as startling to Darwin as it is abhorrent for us today, partly because it represents a total misunderstanding of the theory of natural selection. Though Darwin had placed the preservation of "favored races" in the title of his seminal work, he had always maintained that natural selection operated against individuals. Races—what we view today as varieties or subspecies—did not exist except as a convenient way to group individuals that carried some common traits. Evolution could not act against race, because race was no more than a construct, a fiction. Only individuals existed for the purposes of selection.

Galton reversed Darwin's logic. His work subjugated the individual to the race, and placed the whole of evolution under the control of a Nature that showed its favor by gifting certain races with abilities that outstripped others. By 1883, Darwin was dead, and Galton had a new term for his pursuit of an improved humanity. He called it *eugenics*, from the Greek for "good stock."

In Darwin's absence, the importance of natural selection was increasingly doubted as an agent of evolution. As it had from the beginning, the idea put forward by Darwin and Wallace seemed just too simple. Darwin had maintained that there was grandeur in his view of nature, but how could a theory that could be written on a postcard contain all the glory of the world? Besides, Galton had shown (quite correctly) that Darwin's ideas about the actual mechanism of inheritance were completely wrong. If Darwin was incorrect about how traits were passed on, how could his theory be trusted to predict larger effects of evolution? Darwin's directionless, individual-based ideas had never been a comfortable fit with a hierarchical society.

The idea of a more directed evolution, one that drew from Lamarck and Spencer and Haeckel, found increasing favor, particularly with the public, but even among scientists. Galton explicitly denied that there was any evidence for the kind of inheritance of acquired trait that had been suggested by Lamarck, and in fact took a very strict view of natural selection. However, he also could not accept that evolution was, or should be, directionless.

Galton's idea of race as the unit of evolution fit in well with this more directed approach. If Darwin was wrong, then why shouldn't evolution act against races instead of individuals? If we couldn't see the mechanism for this, it was only because we could not make out the whole of the complex plan. Galton's theories found a natural ally in the growing ranks of Social Darwinists. Where Herbert Spencer taught that government itself was a primitive trait that mankind would soon outgrow with the introduction of a laissez-faire utopia, Galton took it a step further. He maintained that government, by protecting the weak and supporting the "common man," was actually thwarting the progress of racial evolution.

Like Spencer and Haeckel, Galton saw human charity as a brake on the speed of evolution. By helping the poor, weak, and unintelligent, charitable action of all kinds was thwarting the progress of the race. Galton argued that the only way to resolve this problem was for government to reverse its function; to ensure that the best traits were perpetuated, the government needed to take a direct role in human breeding. If Spencer's ideas of evolution drew from medieval ideas such as the Great Chain of Being, Galton's ideas looked even further back to Rome, to Sparta, and to disfigured children thrown from the cliffs of Mount Taygetos.

Today, the eugenics Galton advocated is most associated with the horrors of Nazi social policies, but Galton had plenty of surprising followers at home. The most free-market-centric economists became followers of eugenics, but so did John Maynard Keynes. H. G. Wells was a follower of both Spencer and Galton.

So was George Bernard Shaw. Another follower was a young army officer turned politician named Winston Churchill. By the turn of the twentieth century, the eugenics movement had spread well beyond Great Britain. Alexander Graham Bell, so often remembered as a teacher of the deaf, researched the inheritance of deafness and suggested that deaf couples not be allowed to marry. Woodrow Wilson—who re-segregated the armed forces and was concerned about racial purity—promoted an active program of eugenics and eventually saw the passage of laws in over 30 states that mandated the sterilization of the mentally handicapped.

Galton approved such actions. Like conservatives today, he fretted that poor people were reproducing faster than the wealthy and worried that civilization would be drowned in waves of the underclasses. Galton produced a simple diagram—criminals and the poor at one end, industrial barons at the other end—that correlated wealth and achievement directly with "genetic worth." Just as with Spencer's Social Darwinism, this was a theory that those in power wanted to believe. They were not just richer than the people who worked for them—they were better. Genuinely superior right down to every cell of their bodies.

The sterilization laws were fought by advocates of the mentally ill and physically disabled all the way to the Supreme Court. They lost. In *Buck v. Bell* (1927), the United States Supreme Court held that the State of Virginia was within its rights to sterilize anyone that the state found unfit. More than 64,000 Americans were forcibly sterilized.

American laws based on eugenics didn't stop with the "unfit." Eugenics was also behind the "anti-miscegenation" laws created to block interracial marriages. Laws such as Virginia's Racial Integrity Act blocked interracial marriage in the name of race purity from 1924 until the Supreme Court overturned the law in 1967.

For the American followers of Galton, laissez-faire wasn't just an economic idea, it was a class unto itself. If government's pernicious influence could be removed, powerful "laissez-faires" would rule over the subclass of mediocre humanity. Galton's followers set out to accomplish just that. Eugenics was used as the foundation for executing criminals, for removing children from their parents, and for justifying the extinction of native peoples. The idea spread to Canada and Sweden where thousands more were sterilized, to Australia where it was behind the "assimilation" of mixed-raced children, to China, to Japan . . . and, of course, to Germany.

Galton was not around to see the final stage of his ideas played out across Europe. He died in 1910. He spent his last years working on a fantasy novel about a utopia where eugenics was the foundation of religion, racial purity was valued above all, and the state bred for better humans. He didn't live to see his ideas brought to terrifying life or to see his novel published. His daughter burned the manuscript after his death.

He also didn't live long enough to learn that his ideas not only led to some of the greatest horrors of the twentieth century, but were utterly and completely wrong. Not just morally wrong, but scientifically invalid. His fundamental distortion of Darwin—disparaging the value of individuals—meant that his theory had been established on vapor from the beginning. New genetic studies have shown again and again that race is nothing but an illusion. Race doesn't reflect any significant division of humans. Not only that, the truth is that eugenics was possibly the greatest threat humanity has faced, a threat at least equal to that of the atomic bomb.

To see why, you have only to look at bananas. Big Mike, or Gros Michel—the larger, sweeter, tougher banana that was sold before the 1950s—isn't sold anymore. This perfect banana isn't sold because Big Mike is nearly extinct. A fungal disease struck banana plantations around the world. Infected plants wilted and rotted by the thousands and tens of thousands. Gros Michel was susceptible to the disease. Because all the Gros Michel plants were the same, they were *all* susceptible to the disease. Around the world, the perfect banana died. That funny old song, "Yes, We Have No Bananas"? It records in its lyrics a disaster that destroyed crops, ruined farms, and toppled fortunes. While the Gros Michel plants were being replaced by the Cavendish cultivar, there really were no bananas to be had in many parts of the world.

The Cavendish also has its plagues. A new strain of the same fungus that killed off Big Mike has wiped out Cavendish bananas in several Asian nations. If it reaches the plantations of South and Central America, we may have to revive that old song while banana growers seek out another cultivar to replace the banana you've eaten for the last 50 years.

What makes the bananas so susceptible to worldwide plague is not some problem specific to bananas. What made Big Mike so fragile was low genetic diversity. Other crops, from corn to cows, have demonstrated the same susceptibility. A field full of genetically identical corn may produce more grain than one made up of multiple strains, but it's also far more vulnerable to disease and climate. In evolutionary terms, "pure breeds" are not stronger, they're much more fragile. No matter how powerful the blond-haired Aryans on old Nazi posters may appear as they threaten the *untermensch*, a humanity peopled only by such a narrow group would be a humanity always on the brink of disaster.

Galton read *On the Origin of Species* and was excited by Darwin's description of domesticated animals, but he missed the sheep for the flock. The potential a species holds is measured in the genetic diversity of the individuals that make up that species.

The species itself is nothing but a made-up name. The *individuals* are everything.

Increasingly, modern agriculture has recognized the flimsy nature of our low-genetic-diversity crops. Around the world efforts are under way not only to develop new varieties, but preserve the old. "Heritage tomatoes" are grown in both fields and greenhouses, preserving the flavor that once graced our grandparents' meals. "Heritage chickens" are brought back from the brink of extinction to lay speckled eggs that seem like a novelty to generations of Americans used to pure white eggs from the supermarket. Over the last few centuries, thousands of varieties of domesticated plants and animals have become extinct, and hundreds more are on the brink. It will take a deliberate and ongoing effort to preserve even a fraction of that variety.

Diversity can also be winnowed down in animals and plants that are not domesticated. Cheetahs today have such a low degree of genetic diversity that, in essence, every cheetah is a close relative of every other cheetah. The species has recently gone through a "bottleneck," an instance in which it was reduced to only a few individuals. This makes cheetahs vulnerable not only to disease and to genetic problems, but to a kind of inflexibility in the face of change. Human beings went through a similar bottleneck around 70,000 years ago. All people alive today can trace their ancestry to a small group of humans back then—and be grateful that those people didn't succumb to the problems of low diversity.

When it comes to business, we haven't yet learned the lessons of the banana. Stop at any major intersection along American highways and you're likely to encounter nearly identical sets of retail establishments. The same fast-food outlets. The same big-box retailers. The same gas stations, druggists, and hardware stores. These large, well-funded operations maintained by corporations with headquarters far away may seem more robust than a scattering of local mom-and-pop shops, but that's as much an illusion as the shiny yellow wrapper on Big Mike. Local stores may not have the large bankrolls that their chain-store competitors enjoy, but they're far more flexible when it comes to adapting to the demands of a particular market. It's only the leverage that large corporations are able to bring to bear—which includes using temporary pricing that undercuts all competition—that allows them to replace America's diverse commercial structure with one that's bigger, shinier, and far more fragile. You only have to look at the amount of prime commercial real estate now occupied by *empty* shells that once housed big-box establishments to see that these firms are not immortal, and their deaths can have powerful consequences for a community.

It's always easier just to take the option that looks most attractive at the moment. The road forks, and that neatly traveled-by path draws us in far more readily than the bramble-laden alternative. Galton viewed "mediocre men" as a threat to civilization, but the purpose of civilization isn't to select for the elite: it's to preserve all of us. Civilization is the field in which heritage humanity is sustained. Society is the environment in which far more of us are "fit" in the sense that Darwin actually intended than we would be in the brutal classism of a Social Darwinist "utopia." We don't know what the next crisis to face our species will be, but by maintaining a high level of diversity in all areas, we are better prepared to face that next crisis and survive—both as a species, and as individuals.

321 Jaffa Gate. Jaffa-Thor. Porte de Jaffa.

Jaffa Gate, Jerusalem

G. ERIC AND EDITH MATSON
PHOTOGRAPH COLLECTION, 1908

CHAPTER 9

Camel Puree

Why has not anyone seen that fossils alone gave birth to a theory about the formation of the earth, that without them, no one would have ever dreamed that there were successive epochs in the formation of the globe.

GEORGES CUVIER, *Discourse on the Revolutionary Upheavals on the Surface of the Earth* (1812)

Consider this popular story that concerns a very unpopular gate. In the walls of first-century Jerusalem was a city gate so small that it was of limited use. A man passing through this gate would find it so low that he was forced to bow his head to pass. A tall man might even be driven to his knees. Riding any sort of mount through this gate was out of the question. A camel or donkey led through the tiny opening would find itself pressed from all sides, and any goods normally carried by the animal would have to be removed. So, whether they liked it or not, those using this gate would enter the holy city in a posture of penitence and with their worldly goods—at least temporarily—left behind. The name of the tiny gate was "the needle's eye."

What makes the story so popular is this:

> Then Jesus said to his disciples, "I tell you with certainty, it will be hard for a rich person to get into the kingdom of heaven. Again I tell you, it is easier for a camel to squeeze through the eye of a needle than for a rich person to get into the kingdom of God.[1]

Jesus's statement has been making people wince for two millennia. The idea that "the eye of the needle" might have been the name for a snug gate or mountain pass is comforting to those inclined toward a literal interpretation of Jesus's statement (especially since, despite all the Old Testament smiting, the only people God whacks in the New Testament are a couple who attempt to weasel out of their commitment to hand everything over to the church).

Unfortunately for those seeking to evade a divine poverty edict, tracing back the tale of the little gate shows that the story first appeared in the Middle Ages. It was a kind of sop for those who weren't ready to dispense with all their wealth (but who might have been inclined to share some of that wealth with the established church). Though the story is commonly told today, and has been for centuries, there's no indication that "the eye of the needle" was ever the name for any pass, gate, path, or portal in or around Jerusalem during the first century.

Some scholars have suggested that the object to be placed through the eye of the needle is not a camel at all, but a rope. They argue that this whole dromedary dilemma is based on the mistranslation of a bit of Aramaic, or a transcription error that obliterated the difference between the words for camel and cord. But again, there doesn't seem to be any real evidence that this is so, and similar statements elsewhere in documents from around the same age suggest that *camel* was the word after all.

There's no getting around it, the needle in Jesus's story was of the sewing kind, the camel was of the spitting variety, and the eye was small enough that any camel passing through would have to be first reduced to something very close to its component atoms.

Fortunately for anyone (and any camels) confronting this predicament, it's likely that Jesus was less literal in his interpretation of the statement than many modern followers—stuffing that camel through a needle had been an aphorism for "nearly impossible" for several centuries before some rich kid started sweating about his afterlife. No literal gate existed in the walls of Jerusalem, and no literal needle. The eye of the needle and the nervous camel are only metaphors for the difficulty of reaching sanctity when your heart is more set on wealth.

Likewise, no literal gate sat on the shores of the widening Atlantic at the end of the Cretaceous period, 65 million years ago. But there might as well have been. If this considerably more ancient gate carried a label, it would be this:

IT WILL BE HARD FOR A LARGE CREATURE
TO GET INTO THE TERTIARY PERIOD

We're used to thinking of the end of the Cretaceous as the end of the dinosaurs, and it was.[2] However, the end of the Cretaceous saw not only the end of dinosaurs. The last of the flying pterosaurs, which were unrelated to dinosaurs, also died out. So did the giant reptiles of the sea that Mary Anning had discovered. The shelled ammonites of the oceans—similar to the chambered nautilus, but in many shapes and sizes—vanished. Even among the groups that are still around today, the losses were tremendous. Birds, the dinosaur's direct descendants, might have made it past the line as a group, but many birds didn't. Most birds died right alongside their non-avian dinosaur relatives. Mammals made it—which is surely worth a cheer—but many species of mammals also died at the same time. So did most reptiles. So did most fish. The end of the Cretaceous was a nasty time. Lots, and lots, and lots of things died off that weren't dinosaurs. Overall, about half of all genera and three-quarters of all species failed to make it from one side to the other, from one period to the next.

There was no single characteristic that seems to have been a good predictor of surviving past that ugly line of iridium-enhanced clay in the geologic record, but there was certainly one that was a good indicator of *not* making it: large size. Many small animals died at that time, but *every* land animal bigger than a standard poodle was a goner.

The end of the Cretaceous is far from the only such event. The history of life on Earth is studded with cataclysms that wiped out 10 percent, 50 percent, or even 90 percent of all species.

One of these events came about 185 million years earlier than the death of the dinosaurs, at the end of the Permian period. At that time, the world was dominated not by the dinosaurs but by a group called the *therapsids*. This group, which included mammal-like reptiles and the ancestors of true mammals, ruled over an ecosystem as complex, interesting, and complete as anything that came after. They were a diverse group, with many features that gave them advantages over the other large animals of that period. They ranged from tiny insect eaters to animals weighing over a ton. Spread across the globe for tens of millions of years were dog-like predators, sheep-like plant eaters, and creatures utterly unlike any that followed since. Nearly all of them vanished. Their extinction that didn't just match that which ended the dinosaurs—it was far worse. In a very brief period, "the great dying" saw the end of 83 percent of all genera. In the seas, 96 percent of all species went extinct. Only after this extinction, on a world nearly emptied of life, would dinosaurs start their rise to prominence. Thirty million years passed before the world regained the diversity it lost at that time. As it happens, all mammals today are also therapsids, descended from survivors of that ancient group. The same kind of animals dominated the land both before and after the dinosaurs, with a pair of major extinction events neatly bookending the period of dinosaur rule.

Each of these extinction events produced losses all across the spectrum, but large creatures almost uniformly vanished. It was like a sign from a carnival ride in reverse—only creatures shorter than this line can cross into the next age. Even during the time that humans have been around, "mega fauna" have experienced widespread extinction. Humans were the likely cause of at least some of the most recent losses, but not all. These extinction events seem to have a variety of root causes: asteroids, volcanoes, people. What they all have in common is the problem that big creatures have in surviving. No matter what puts the stress on the environment, the big stuff is the first to fall.

As far as our understanding of nature goes, that's a good thing. Through the eighteenth century, very few people recognized that extinction had happened at all. Thomas Jefferson had received specimens of several giant North American animals from a site in Kentucky, and when he dispatched Lewis and Clark into the West, he did so with confidence that they would return with tales of living mastodons and ground sloths. That didn't happen. As more and more Ice-Age giants and the bones of dinosaurs were pulled from the earth, the problem became increasingly clear—many things that had lived on the planet in the past were no longer around. Jefferson might not have given up the search, but over in Europe one man had put his finger on a very large problem.

By 1799, the great French anatomist Georges Cuvier had published two papers on elephant species and in his review had included species known only from fossil bones. The vanished elephants were distinct enough from those still living in Africa and India that nobody doubted they were different species. They were also large enough that their absence from the landscape was noticeable. If extinction had worked the other way, if it had tended to take out things as small as mice but leave larger creatures intact, it might have been much longer before anyone noticed that the animals of the fossil record hadn't simply gone to live elsewhere but were entirely gone. Because the gates of the extinction events so often closed in the faces of large creatures, the change in fauna was very clear even after only a few decades of organized fossil collection. Many, if not most, of the animals that had once lived on Earth were no longer around.

And all those extinct animals were a problem. Looking around the world at the time, Cuvier and others could see that the planet didn't seem to be exactly missing all these vanished giants. They were gone, but their absence didn't leave a hole. Where mammoths had cropped the grass of Europe, other animals were taking care of that in modern times. Most of what was found in the rocks represented species that were extinct, and yet the world wasn't lacking for diversity. The fossil record revealed that the great majority of creatures that had once lived were no longer to be found. Instead new creatures had appeared. The populations of animals that lived in any area were far from constant.

To explain the turnover of species, Cuvier embraced the geologic school of catastrophism. At the time, most catastrophists had a simple explanation for why so many animals were extinct: they had missed the boat. That is, they were left behind during Noah's flood. Though Cuvier initially wrote extinction off to some single past event, like the flood, he was never one for point-ing at Noah. In fact, he later modified his approach to accommo-date a number of different catastrophes. So Cuvier understood that most fossil animals were extinct, and he had formed a revised theory of catastrophism that was closer to the modern under-standing than most others of his time. Still, there was one thing he didn't accept—evolution.

Part of the reason for his denial was personal. The prevailing theory of evolution at the start of the nineteenth century was one that had been put forward by Cuvier's countryman, Jean-Baptiste Lamarck, and Cuvier did not like Lamarck. Unlike Lamarck, Cuvier was not an aristocrat. He was the working-class son of a retired military officer who, through extremely hard work, diligence, and a frightening level of genius, had risen to be quite possibly the most respected scientist in the world. Lamarck's theory—which even in the early 1800s was being used to pro-mote the sort of "natural superiority of the ruling class" that would later be associated with Social Darwinism—left Cuvier cold. Additionally, Cuvier was reluctant to sign on because the theories of evolution then put forward depended on the one thing Cuvier didn't see in the fossil record—slow, uniform changes.

A hundred and forty years after Cuvier's death in 1832, Niles Eldredge and Stephen Jay Gould published a paper that was very much in agreement with what Cuvier had seen. They maintained that evolution didn't move by an unbroken series of tiny changes across a broad population. Instead, the fossil record indicated that change most often happened to small, isolated populations over relatively short periods. For millions of years at a time, spe-cies survived nearly unchanged, only to be replaced by new spe-cies during periods of rapid turnover. Eldredge and Gould named this pattern *punctuated equilibrium.*

This pattern had been noticed for a long period—in fact, it was noticed by Darwin and his contemporaries. Some assumed it was a measure of the incompleteness of the fossil record. Eldredge and Gould said no, the record wasn't at fault; the idea that change happened at a steady pace across a whole population was simply a bad assumption.

Eldredge and Gould didn't mean that evolution proceeded in large, abrupt changes. All evolution happens gradually. But the pace of evolution is not steady, and—because the period during which a creature is in transition can be short—the fossil record tends to show the stops on the evolutionary bus route far more often than it shows the trips in between. The answer to what had shaped the history of life was not the slow steady change predicted by uniformitarianism or the chaos implied in catastrophism. A compromise of the two was needed to explain what was recorded in the rocks.

Few equilibriums are as "punctuated" as the times around a mass extinction. Such events not only represent a sudden downshift in life's diversity, they also mark the start of a mad scramble to refill the void. The "great dying" at the end of the Permian was also the starting gun for the expansion of dinosaurs and other archosaurs. The end of the Cretaceous wrapped up the story of the dinosaurs and started us on the path to the world we see today. Extinctions aren't just about failure, they're about opportunity.

This pattern—slow and steady alternating with periods of stress and change—is present at all levels of the natural world, and each part of the pattern has its role in determining the course of life. A world of nothing but one disaster after another would likely be a world all but empty, and life would be driven down to only the smallest and hardiest of survivors. In fact, the history of

Earth includes several events of such magnitude that they came very close to a complete "do-over" for life. On the other hand, a world of utter stability might be one of stagnation, where very similar organisms persist for billions of years. Ironically, such a world might also be populated with nothing but the simplest forms of life. Without the periods of stress and extinction, complex life might never have won a foothold. We would not be here without both parts of the evolutionary cycle.

In our own lives, we may try to avoid disaster (even if it provides a little variety), but the complex systems around us rarely allow us to experience smooth sailing for very long. In very few places on the planet can you live out a lifetime without experiencing political turmoil, if not outright war; in fewer still can you pass through your life without massive restructuring of the economic order.

For the last two centuries, as the industrial age has played out and moved into the information age, the world has experienced the erratic swings of the "business cycle." These cycles mimic the natural world in many ways, including one that many people find peculiar—it's the periods of disaster or near disaster that generate most new companies. The reason for this isn't particularly pleasant. People start new businesses in bad times simply because they've lost existing jobs or are unable to find employment at an existing company. Similar to species that appear during the stressful time after a mass extinction, the long-term survival rate of these new businesses isn't particularly good. But it doesn't have to be. With enough raw material, the winners will emerge, grow, and become the giants of a future cycle.

The reason giants go tumbling when stressful times come along is that giants are children of stability. Just as with catastrophic events in the natural world, times of stress can bring on business failures at all levels, but these periods are particularly tough on large companies. The rationale isn't hard to understand. Not unlike large animals that require more resources to live and so are susceptible to changes in their environment, large businesses exist at the center of complex social, political, and economic webs where a failure in any strand can spell insolvency. You don't form these webs in the middle of disaster. They take time and stability to evolve.

Post–extinction event worlds are worlds for small animals. Post–economic disaster worlds are worlds of opportunity for small businesses.

When conditions are steady over a long period, they can promote the development of very large creatures capable of utilizing all the resources of their environment very efficiently, in ways that are difficult for smaller organisms to emulate. Stable economies support the creation of massive corporations, which are capable of exploiting every aspect of law, trade policy, differential pricing, and global resources.

In the case of banking, the combination of deregulation and nearly two decades of steady growth created "monsters" that grew ever more dependent on an incredible array of exotic securities. Those involved may have felt that they were responding to a fast-moving environment, but it wasn't speed that promoted the creation of credit-default swaps, intricately recursive investments, and world-sized stacks of collateralized debt obligations. Stagnation instigated these instruments. Only in a system held in perfect balance can such delicately embroidered creations appear. The merest breath of the economic wind is enough to rend such structures like damp tissue in a storm.

Of course, unlike in the natural world, many of our business giants did not die in the latest cycle of disaster. Because we didn't let them die. To some extent, this represents a new "evolutionary path" in business. It's not that these banks are "too big to fail." In both nature and economies, big is a precursor to failure. Big makes it easy to fail. However, these companies have extended their reach into the political sphere to such an extent that they have an alternate source of revenue. As big organisms do everywhere, they're exploiting their resources in a way unavailable to smaller competitors. Elephants unable to reach branches may simply uproot the tree. Giant banks unable to draw enough sustenance from the economic sphere instead insert a feeding tube into the political sphere and suck in the nutrition there. They have become "fit" in their environment in a whole new way.

This survival by bail-out may work. It may even be a good thing. Disasters might be times of opportunity, but they are first and foremost, well, *disasters*. Keeping alive the current crop of giants may help to blunt the "extinction" of many businesses other than just those getting direct support.

But one thing is sure. By holding up our injured giants, we're locking up the resources that would otherwise go to create a new crop of small businesses. The crash of a massive tree in the forest is a tragedy to the organisms that live in that tree, but it creates the rich soil in which the next generation will sprout. The new businesses that might have formed had we allowed the banks to fail would themselves have formed a new generation of giants, one more attuned to current business conditions. Instead, we have our old giants—only fatter, and not particularly smarter. In future business cycles, the absence of a new generation may well be critical.

Steller's Sea Cow

EXTINCT MONSTERS, REV. H. N. HUTCHINSON, 1883

CHAPTER 10

Late to the Party

My scientific studies have afforded me great gratification; and I am convinced that it will not be long before the whole world acknowledges the results of my work.

GREGOR MENDEL, correspondence (1883)

Just a few miles down the road from my home is a "home improvement" store. It's a large, low building flanked on one side by a place for plants and garden supplies and on the other end by a chain-linked bit of asphalt that used to hold lawn tractors and gas grills. I say "used to" because the store is empty.

It was constructed in 1995 as a Builders Square. When it opened, the little Ace Hardware down the street and a local garden center next door went out of business. Only two years after it opened, the blue and gold markings of Builders Square were replaced by the green "HQ" of Home Quarters Warehouse. It stayed that way for a couple of years. Then an even larger store opened a few blocks away, this time with the orange blaze of Home Depot. Within months, the green HQ store was closed.

Now the orange store has been flanked by an even bigger, newer store, bearing the blue markings of Lowe's. The orange store is starting to look a little shabby by comparison. Ace to Builders Square to HQ to Home Depot, and maybe now to Lowe's. Surely there's someone with another logo—purple or yellow or maybe chartreuse—lurking on the horizon.

This commercial theater of dancing colors, played out on a few blocks of one street near St. Louis, is just an expression of a much larger drama. What happened in this competition and succession of businesses was guided at least as much by events that occurred at corporate headquarters hundreds of miles away, and by other struggles that took place in other cities, as it was by competition along that particular street. But this kind of succession, in which apparently successful establishments are swept aside by new entries, is a common feature of the business world. The principle is often called the *late mover advantage*.

Big players in an area of business can seem unassailable. They already hold the niche. In fact, their names may be so associated with a certain line that it's hard to think of that industry without thinking of them. Xerox equals copy machines. IBM equals computers. Sears equals retail.

But Goliaths fall to scrappy upstarts every day. With every passing year, stores and fixtures that were once new become more out of date and in need of repair. Researchers once at the cutting edge turn into the unadventurous establishment. Stores built in locations that were once prime find themselves moored far from the new development and the new establishments that draw in shoppers. Unsold inventory accumulates. Groundbreaking patents expire. Staff enthusiasm wanes. Although an ever spreading web of stores means more opportunity, it also magnifies the cost of each mistake. All the while, the late movers watch from the wings, noting each error, searching for weakness, looking to copy what works and discard what doesn't, and insert themselves into the market at the time of maximum advantage.

Even those businesses that have been around for decades can suddenly find themselves in a struggle for their lives against competitors that didn't exist until recently. Ask Woolworth's. Ask General Motors.

Ask sea cows.

In the natural world, this sort of pattern is common. A role in an environment can be dominated for a long period by a single species, or by a handful of closely related species. This dominance can last millions of years and spread over huge areas. But visit that same environment in the same location only a few million years later, and you may find a quite different creature filling the same role.

When Charles Darwin published *On the Origin of Species* in 1859, the fossil record was like a dark room with a few scattered spots of dim light. Fossils had been collected from only a few regions, there had been little systematic collection, and much of what had been found wasn't clearly identified. Given the murkiness of the situation, it wasn't surprising that Darwin's critics, such as anatomist Richard Owen, were able to assert that the fossil record didn't show the type of gradual transitions Darwin had predicted. Darwin agreed. In fact, Darwin worried that the fossil record would never be clear enough to act as evidence of natural selection. Even when the rocks yielded a specimen that was clearly an intermediate between two other forms, that specimen itself seemed isolated.

Critics pounced on Darwin's theory by pointing out this lack of evidence for gradual transition in the fossil record as a weakness. Even as time passed and the number of fossil species rose from the hundreds to the hundreds of thousands, skeptics still huffed that there was "no evolution in the rocks." A great deal of their reluctance was simply a form of *reductio ad absurdum*. Presented with a form that was at the midpoint between two species, the critics of evolution would shriek about the absence of a specimen three-quarters of the way toward one species and one-quarter of the way toward the other. If something turned up that satisfied that complaint, they'd move to arguing over seven-eighths. No matter how fine a sequence is presented, there is always some gap that determined reductionists can use as a wedge for their complaint. This method turns every new discovery into a win for the doubters because every new piece in the puzzle only creates more missing pieces. It worked in 1860, and it's still a mainstay of anti-science creationists.

Darwin may have doubted, but most of his supporters over the last century and a half have assumed that with sufficient exploration many of the gaps in the fossil record would be filled, and the predicted pattern of gradual transformation would be confirmed. That was certainly true in many respects, as more and more fossils confirmed exactly the sort of large-scale transitions that had been predicted in *Origin*. But on a smaller scale, the gaps stubbornly refused to close. At many points in the fossil record, the transitions from one species to the next remained oddly abrupt, without evidence of transitional forms.

It was such a serious problem that even scientists who believed in evolution began to doubt Darwin's ideas about the way in which evolution operated. Various forms of Lamarckism—not just the versions provided by Spencer and Haeckel—competed with natural selection during Darwin's lifetime, and they continued to attract followers in the decades that followed. When Darwin died in 1882, this alternate view of evolution experienced a kind of resurgence. Increasingly, natural selection seemed to be caught in a trap. The theory seemed inadequate to the task of explaining how the variations in species arose, or how traits were passed along, or how multiple species developed from a single source.

Darwin himself was deeply aware of the deficiency of his

ideas in explaining the how of change. He turned to an idea called *pangenesis* as a means of explanation. In this theory, each cell in the body produces a kind of local blueprint that is shipped down to the reproductive hot zone for exchange during sex. The presence or absence of these little capsules could affect the offspring. No real evidence supported the theory. Worse, what evidence there was increasingly argued *against* pangenesis. The idea only opened a host of new questions, it provided few satisfactory answers, and its predictions about inherited features just didn't seem to match real-world results. When tests were carried out, pangenesis flunked. This gaping hole in the side of natural selection was another cause for the reduced acceptance of Darwin's ideas in the years immediately following his death.

However, the experimental evidence that Darwin needed to bolster his work had already been collected and published decades earlier. In the period from 1856 to 1863 (neatly framing *Origin's* publication at the end of 1959), Gregor Mendel, a priest working in Austria, cultivated nearly 30,000 pea plants. Growing that number of plants isn't unusual (modern farmers plant as many as 125,000 soybean plants on a single acre to maximize yield), but most people plant fields of uniform seed, whereas Mendel was carefully cross-breeding different varieties and noting the resulting offspring. Close observation of traits over generations of plants allowed Mendel to determine the way in which inheritance worked through dominant and recessive alleles long before the genes themselves could be detected. He published his paradigm-altering observations in 1866 and sat back to . . . deafening silence.

Although the delivery of the first Darwin-Wallace paper on natural selection drew little enthusiasm when it was read for the Linnean Society in 1858, that paper still received more attention than Mendel gathered when he read his work on inheritance to the Natural History Society of Brünn. The paper, published in an obscure Moravian journal, languished for decades. Sadly, Darwin never became aware of Mendel's discoveries and Mendel never experienced the acclaim his work would eventually earn.

Not until the start of the twentieth century (a few years after his death) was Mendel's work rediscovered and his experimental results replicated. Not only did the work revolutionize biology almost overnight, but it also provided the key to understanding the ability of organisms to change while preserving traits from the parents. Both abilities were required by Darwin's theory of natural selection. Over the next three decades, a growing under-standing of genetics and population dynamics would lend ever-increasing support to Darwin's work. More evidence would come from every field of biology and paleontology, as strand after strand of supporting theory came together. The fossil record might not demonstrate gradual change, but year by year, paper by paper, the study of biology was gradually transformed from multiple, disconnected areas into a cohesive unified discipline—with Dar-win's work right at the center.

The major complaints of Darwin's critics had been solved, but that vexing problem of the gaps in the fossil record—gaps that refused to go away no matter how many specimens were col-lected—still remained. In 1942, evolutionary biologist Ernst Mayr made a large contribution to solving the problem. According to Mayr, the gaps didn't exist because organisms changed abruptly. They changed gradually, just as Darwin (and by then, genetics) predicted. But not all examples of a species were represented in the fossil record and they were not all taking the same genetic trip. Local populations—particularly small, isolated populations—could diverge from other examples of a species. The local popula-tion of dust bunnies remained bunnies in the living room, while another group of bunnies were slowly becoming dust hares behind the barrier of the bedroom door. When the hares—who had been gradually evolving offstage—came tumbling out to replace their relatives, watchers in the living room may have thought they were witnessing a rapid coup. But truthfully, the genetic gearbox of change had not been running in overdrive. In this way, you could get not only apparently abrupt change but develop a new species at the same time that the old species was still around. You could also get multiple species from a single stock.

By the early 1940s, the modern evolutionary synthesis was in place. It can be summed up in five simple statements:

1. All evolution is consistent with genetic mechanisms.
2. Evolution takes place gradually, through small changes.
3. Natural selection is the overwhelming force driving evolution.
4. Genetic diversity within a species is the raw material against which selection takes place.
5. The rate of change is not constant, but all change is gradual.

The explanation of abrupt changes in the fossil record was given a more thorough examination when Eldredge and Gould published their paper on punctuated equilibrium. Like Mayr, they acknowledged the role of isolated populations in driving evolution. Under punctuated equilibrium, if you were to peel back the layers of a particular bed of fossil-bearing rock so that you were observing the presence of one species over time, you would not find a series in which type A creatures gradually add more characteristics of type B until the original species has been replaced. Instead you find A. Then A. Then A again . . . right up until the point where you find B (with, perhaps, A still living in the same area). It was this pattern—a long period of species stability followed by relatively abrupt change—that Gould detected in his lengthy studies of land snails and Eldredge saw in his fossils of trilobites.

Punctuated equilibrium has often been touted by critics as proof that Darwin was "wrong," or that there's some great controversy within the scientific community about the mechanism of evolution. Neither is true. Punctuated equilibrium doesn't overturn Darwin's theory in any sense—though it does confound those followers of Darwin who expected to find slow, steady processes everywhere in the fossil record. Some critics have even tried to assert that punctuated equilibrium provides some proof for a kind of ongoing creation in which new creatures are dropped into the mix from time to time. That's even further from the mark.

A few of those who looked at punctuated equilibrium had contended (and Gould was prone to agree) that evolution didn't always take gradual steps, and that punctuated equilibrium at least occasionally recorded "hopeful monsters," meaning a large genetic shift might take only one or two generations. From the time of Darwin, people were reluctant to believe that features such as a bird's wings could develop gradually, and the hopeful monster theory presented an out. You didn't have to wait for a dinosaur to grow a wing piecemeal, if some lucky freak could take a big jump along the path.

But this idea proved to be wrong. Recent genetic analysis backed up what Darwin had said all along, and it was restated in the modern synthesis of the 1940s. All evolution takes place gradually. Happily enough, Darwin's fear that the fossil record would never be strong enough to provide evidence for natural selection has proven untrue. The fossil record is rife with evidence of changes that took place both slowly and relatively quickly.

When evolutionary biologists talk about relatively fast change, the key word is *relatively*. A change that looks sudden in the fossil record can still be quite gradual on most scales. Say, as an example, that human beings were to trip along for the next million years with a height of around two meters tall, but then future exo-paleontologists find human beings averaging three meters tall only a million years later. That's quite a significant and sudden change on a geologic scale—a switch from David to Goliath. But a million years is time enough for 50,000 generations of humans to come and go. In that million years, each generation would need to grow an average of only two-hundredths of a millimeter taller than the previous generation to accomplish this "abrupt" change. Your height changes far more than that when you put on your socks.

Punctuated equilibrium doesn't discredit Darwin's idea of gradual change. It doesn't require hopeful monsters (one of my favorite phrases in all of science) to drive the progress. By any rational measure, the change from one generation to the next is quite small. What punctuated equilibrium did was simply acknowledge that the general state of change was miniscule and that most organisms stayed more or less the same for long periods. This long-term stability makes the periods of relatively greater change look rapid by comparison.

That's not to say that the slow, wholesale change of a widely established species can't occur. A few very good examples of exactly this kind of evolution (known as *phyletic gradualism*) are preserved in the fossil record. One of the best is found among the fossil manatees along the Pacific Coast of North America.

Sirens (the group that includes manatees and dugongs) arrived along America's West Coast during a tropical period in which the waters off the coast from California to Canada supported warm, grass-filled bays more like those found along the Gulf Coast than anything seen in the Northern Pacific today. In that environment, manatees prospered, and several species sometimes shared the same area. Then, around seven million years after the manatees had arrived, the tropical climate abruptly ended. Water temperatures cooled. A period of mountain-building drove the land up and turned gradually sloping shorelines into sharp cliffs. Manatee-friendly bays that had been filled with sea grass turned into deep, cold water trenches. Most of the Pacific manatee species died out, but one of the smallest, *Dusisiren*, managed to hang on.

Gradually, through a series of slow transitions that would have greatly pleased Darwin, *Dusisiren* increased in size and became better adapted for eating the new kelp beds that were then replacing the grass. Probably as a response to the insulation needed in colder water, the descendants of *Dusisiren*, now different enough to be considered a new species called *Hydrodamalis*, grew much larger than the manatees and dugongs found in tropical and subtropical environments. These sirens spread northward and westward along the coast of Alaska to the islands west of there. As they grew bigger, they gradually lost the tusks and specialized teeth other manatees use to uproot and eat grass. Their forelimbs changed from standard manatee flippers into tough little "arms" capable of holding the heavy creatures steady against pounding waves, or even "walking" over submerged ledges in search of kelp. All these changes seem to have happened very slowly and also quite uniformly across the population. By the early Pliocene, around 5 million years ago, big toothless *Hydrodamalis* had completely replaced his smaller forebears—and was still growing.

When European explorers unexpectedly encountered the creatures among a series of small islands off the coast of Russia, the last of the Pacific sirens had become giants. They reached lengths of more than 8 meters (around 26 feet) and weighed in around 10 tons—a size that made them as long as the largest killer whales, and about three times as heavy. They formed a floating colony around the shore of Arctic islands, and watched the explorers with no apparent alarm.

If you're unfamiliar with the idea that there are such creatures in the Pacific, there's a very good reason. Less than 30 years after Steller's sea cows (named after their discoverer) were first encountered in 1741, the last one was killed and eaten by traders on their way from Russia to Alaska.

Manatees survived over 25 million years of changing conditions in the Pacific, but they didn't survive human contact.

Back in the world of business, phyletic gradualism can seem as rare as it does in the fossil record. Though fighting off the pressure of a "freshly evolved" upstart can be difficult for established corporations, it does happen. *First* movers have their own set of advantages. Those advantages are particularly evident in technological fields where limited resources (often limited *human* resources) and a specialized knowledge base can give the first entrant into a new area a virtual lock on the field while competitors struggle to break in.

In more general commerce, early entrants don't enjoy that same edge, but they do have a foundation of pure experience that can sometimes turn the tide. A few miles in the opposite direction of my hardware parade is a small town where Wal-Mart moved in and existing local stores began to fail. But rather than folding its tent, Sears decided to make a different kind of play. It opened up a "Super Sears" on the scale of the Wal-Mart but with a different mix of goods. Whether this punctuation of Sears's long equilibrium will be effective is hard to say, but it's interesting to watch.

Wal-Mart itself is undergoing a kind of phyletic gradualism, growing from merely Brobdingnagian to an absolutely Galacticusian scale. Maybe that change will serve to keep the giant ahead of the circling Davids, but I wouldn't place a bet on it. Over the long term, evolution is particularly unkind to giants. Only 27 years after their discovery, someone ate the last Steller's sea cow. No matter how powerful it appears at the moment, it would be risky to bet that any retail giant would survive much longer.

Wilderness Road

LIBRARY OF CONGRESS, 1755

CHAPTER 11

If You Go Out in the Woods Today

This is the forest primeval. The murmuring pines and
the hemlocks,
Bearded with moss, and in garments green, indistinct in
the twilight . . .

HENRY WADSWORTH LONGFELLOW,
Evangeline (1847)

The next time you're upset over how long the highway department
is taking to complete that always-in-the-works intersection, you can
at least be grateful that the project didn't take 30 years. Of course,
if that road construction was not just transportation, but stimulus
package, engineering experiment, economic policy, and immigra-
tion plan all rolled into one, you might be a little more tolerant.

Construction on the Cumberland Road didn't begin until
1811, but the idea for it originated in an argument that took place
in George Washington's office in 1790. In the face of a fiscal crisis, Alexander Hamilton was ready with a plan for the Ameri-

can economy. Thomas Jefferson was ready to stop him. After hearing the details, Jefferson called Hamilton's plan both illegal and immoral, and he quickly rattled off a half dozen principles imperiled by this proposal. Hamilton rose to its defense. The two men had sparred on any number of occasions and sometimes seemed to look on their arguments as a sport. Washington was not amused; he put a stop to Jefferson's protests and awarded the match to Hamilton. Of the two, Washington liked Hamilton better. It was hard to blame him—for all his good qualities, Jefferson could also be something of an ass.

Washington's administration was then facing a plethora of problems. Nearly all of the young nation's manufactured goods were imported, and domestic manufacturing was finding it hard to get started when cheap goods were pouring in from overseas. American industry threatened to be trapped as a supplier of raw materials for companies located elsewhere. Hundreds of state and local banks were operating under different rules, different rates, and even different currencies. Increasing population was driving up demand for land and produce in the East, while goods and agricultural products in the new western settlements had no access to eastern markets. The country was awash in speculators taking advantage of differences in exchange rates and shipping costs to rack up fortunes while farmers and craftsmen suffered.

The ideas that Hamilton introduced to address these issues were dramatic then and would be more shocking if they were proposed today. First, Hamilton argued that the United States needed to impose high tariffs to make imports less attractive. Cloth, furniture, all the various manufactured goods coming into American ports from European makers would be subject to as much as a 25-percent surcharge. These higher prices would provide a margin for American industry to expand. Later generations might associate such measures with unproductive protectionism and even view tariffs as a trigger for economic depression. But the first American administration (and many that came after) did not agree. In addition to the tariffs imposed on imports, subsidies would be awarded to American manufacturers.

Another of Hamilton's proposals was that the country needed to focus on infrastructure, like the Cumberland Road, as a means of providing access for settlers and transport for goods. Roads and canals were planned across the country. These "shovel-ready projects" (which at that time were often carried out by small armies of men actually carrying shovels) vied for funding and priority just as high-speed rail systems and interstate highways do today. At the same time, the federal government would boost the states by taking over the debt they had incurred during the Revolutionary War.

However, there was a problem with the list of projects waiting their turn in Congress: funding. The new government had no way to pay for all the proposals that quickly came to cover Washington's desk.

Hamilton had an answer for that issue, one that has its own echoes in modern times. Hamilton proposed that the US government should become the largest stockholder in a bank. Not an existing bank, a new bank. Hamilton's idea was that the government would create this bank as a private company but would own the first $2 million in stock. The government would also lay down the rules for the bank, including limits on who could buy stock, what type of investments the bank could make, and how directors would be chosen. This bank would then lend the government the money it needed to handle its infrastructure projects. There was only one major obstacle—the US government didn't have anything close to $2 million available. That was where Hamilton got really clever. The bank itself would lend the government the dollars to buy the stock in the bank that the government was creating. Got that? The government would begin paying back that loan at the end of 1791, but even those payments were more than the government could afford. To solve that issue, Hamilton made a proposal popular with politicians in any age, a "sin tax." In this case, the tax was on imported and domestic spirits (to see how well that went, just look up the Whiskey Rebellion).

Those who worry that we've wandered too far from the vision of the founding fathers might want to remember that in George Washington's first term, the government was involved in selectively restricting imports and paying out subsidies, bailing states out of debt, creating private companies, being the major stockholder in a bank, and executing a "stimulus plan" of infrastructure projects. All paid for by new taxes.

The Cumberland Road was one of the largest of the projects envisioned by the government. When it broke ground in 1811, it aimed to stretch from the Potomac to the Ohio, and it wouldn't be just a rutted track for horses and wagons. It was the first major road in the nation to employ the new system of macadam surface—a slurry of crushed rock still used on roads today. This was an expressway into America's interior.

By the summer of 1818, the road-building crew had entered the foothills of the Appalachian Mountains, and these men were cutting their way into the forest. The woods they experienced were massive, dark, boundless . . . terrifying. Though centuries of settlement had pecked away at the woods to the east, and more and more Americans were now moving to the other side of the mountains, the deep woods were still feared. For the men chopping their way through that forest and laying a road with nothing but hand tools, the forest must have seemed endless. And a royal pain in the neck.

More than a century later, some still felt that way. In the wood yard at Rancho del Cielo, Ronald Reagan was famous for swinging his ax to split piles of kindling. Chopping logs may have kept

Reagan fit, but when it came to the source of that wood, he was just as famously unconcerned. During the California governor's race in 1966, candidate Ronald Reagan made his position clear on a proposed national park to preserve ancient trees.

> I think, too, that we've got to recognize that where the pres-
> ervation of a natural resource like the redwoods is concerned,
> that there is a common sense limit. I mean, if you've looked
> at a hundred thousand acres or so of trees—you know, a tree
> is a tree, how many more do you need to look at?[1]

On another occasion, Reagan was asked about the beauty of a grove of 2000-year-old redwoods.

> I saw them. There's nothing beautiful about them. They're just a
> little higher than the others.[2]

In the end, Redwood National Park did not contain "hundreds of thousands of acres" of the giant trees. The park totaled 58,000 acres—including roads and parking lots. In exchange for giving up just 5000 acres of trees that were marked for logging next to the park, the timber industry was given 13,000 acres of old trees elsewhere. Those old trees are now gone. A single giant redwood—300 feet tall and 26 feet in diameter—can yield almost half a million board feet of lumber. If the other 8000 acres had been immediately set aside, we could replace the trees that were lost—in another 2000 years.

At least those who live along the West Coast still have some places they can see an approximation of the forest as it once was. If you squint just right, if you pretend not to notice the foot paths and informational signs, you can see something of what must have confronted the settlers whose wagon trains came out of the plains and desert into the coastal forests of the West.

For those to the east, no such time capsule is available. The forest that those road crews saw in 1811, and that earlier settlers saw as they landed on the Atlantic Coast, is lost. Absolutely and utterly gone. Even in the most remote stretches of Maine or the most pristine areas of the mountains, what we have today is a forest so different from the woods of only a couple of centuries ago that it gives little sense of what conditions were like back then. In fact, when modern scientists looked at sketches made by the first explorers who sailed into eastern American rivers and hiked into those forests, they routinely thought those early explorers had exaggerated. No one could believe that the environment had once been so rich, so diverse, so alien to today's setting, and above all so blanketed in giants.

The biggest reason for the change is the simplest one: time. Trees live for a long, long time. And the largest trees tend to be the oldest. From the sixteenth century on, eastern white pines were harvested from America's East Coast. Reaching up to 230 feet tall, these trees were perfect for making the masts of large sailing ships. Barges were sent up rivers, and well before there were lumber mills, navy crews ranged across the countryside bringing out dozens of these giants at a time. These trees were over 500 years old. The tallest tree in New England today is also an eastern white pine. At nearly 170 feet, it's an imposing tree, but it stands alone. It will be centuries yet before that tree matches up to the ones that once existed in vast groves, and it would take more centuries for neighboring pines to reach such heights.

The same factor of time affects other trees of the east. When life spans are measured in centuries, even logging that happened in Colonial times still stunts the forests today. But another factor weighs even more heavily in the difference between what the forests were and what they now are. Starting in the first half of the twentieth century, diseases have remodeled the forest beyond recognition. The elegant American elm, whose wave-shaped boughs lined streets throughout the nation, was ravaged by Dutch elm disease. This tree has not vanished, but most individuals now die well before they reach their first century and long before they reach the impressive height and girth that once endeared them to both city planners and poets. In many towns the sign announcing "Elm Street" is all that remains to mark avenues once shaded by these giants.

The loss of elms is disheartening, but nothing has reshaped the forest like the disappearance of the most iconic of Eastern trees, the American chestnut. The chestnut did not reach the height of the great eastern pines, but no other tree had the colossal canopy, the massive limbs, the overwhelming *presence* of the chestnut. It was a fast-growing tree, good for fence posts and railroad ties in only a couple of decades—and for furniture, fiddles, homes, barns, and ships. Later it would be used for telephone poles. The wood resisted rot, the tree tolerated a range of soils and weather, a single adult tree could spread its boughs over half an acre, and it produced a bounty of nuts. It was one of America's greatest resources.

When the Bronx Zoo noticed an illness spreading among its chestnuts in 1904, one in four broadleaf trees in the forests of the Eastern United States was an American chestnut. Forty years later, perhaps 100 large trees remained. Some trees might have been resistant to blight among those forests at the time, but as the giants began to die, loggers moved in and worked furiously to take what trees remained. If naturally resistant trees did exist, they were turned into someone's dining room table or roof shingles.

With the elimination of the chestnuts, the nature of the forest radically changed. This kind of dramatic transition is recorded many times in the fossil record. It doesn't take an Earth-shaking disaster or a massive change of climate to radically alter an environment. A species such as the chestnut provides homes and food for dozens, if not hundreds, of species of animals and acts as a kind of regulator on the growth of other plants. The loss of such a species throws the ecology out of balance. Plants and animals faced drastic changes because of its loss and struggled to locate new sources of food, new locations for nests, new positions in the forest hierarchy.

In many ways this change mimics the kind of change that must have happened at many of the large extinction events in the past. When dinosaurs were gone from the scene, it was a disaster for those animals that depended on the dinosaurs for their place in the environment (and there must have been many), but for those who could survive without their giant neighbors, new possibilities were opened up. Though it seems very likely that the asteroid that struck Earth at the end of the Cretaceous was the crowning event in the extinction of the dinosaurs, it may not have been the only cause. Paleontologist Robert Bakker has suggested that dinosaurs might have been done in by the same cause that killed the chestnuts—foreign diseases.

At various times during the Mesozoic era, the continents were connected by land bridges or separated by swathes of open sea. Dinosaurs that evolved in Asia—like the ancestors of tyrannosaurs—passed across land bridges to North America. Giant sauropods lumbered from the Americas to Africa. As dinosaurs moved between different regions, they brought with them diseases that had developed in isolation and introduced them to areas where there was little immunity. However, this kind of mechanism would be unlikely to result in widespread extinction.

Among humans, European visitors brought diseases to the New World that came close to wiping out many native populations, but Europeans were able to spread such virulent diseases because they were members of the same species and resistant to the diseases they carried through generations of exposure. The blight that killed the American chestnut has little effect on Asian trees of a related species. So even if an out-of-town *T. rex* wandered in and introduced a nasty bug to its local relatives, it wouldn't lead to complete extinction—just a replacement of the locals by resistant invaders. Invasive disease is unlikely to be the cause of general extinction, but when organisms from long-separated ecosystems collide, it can certainly lead to a local disaster. Some of the punctuated equilibriums we see in the fossil record may well be caused by disease helping to displace a species.

Beneath the boughs of the chestnut, it certainly was the twilight world of Longfellow's poem. The nuts were not only a major source of food for the forest creatures; they provided both food and income for the people who lived in the shadow of the great trees. The chestnut's range stretched all along the path of that Cumberland Road—from the East Coast to the Ohio Valley, as well as north into Canada and south nearly to the Gulf of Mexico. The loss of the trees was not only an ecological disaster that refigured the landscape but an economic loss that exceeded that of the loss of any industry. When the chestnuts were gone—4 billion huge trees spread over 200 million acres—America was a different place.

Nowhere in the United States could you see what those men saw cutting their way through in 1811. To get some sense of what the forest looked like, you'd need to travel to a park that straddles the border of Poland and Belarus. The Białowieża Forest was owned by the kings of Poland from the fifteenth century and made into a hunting preserve from the sixteenth century on. A few small villages were built in the woods over the centuries, and German forces set up a lumber mill during World War I, but much of the forest remains essentially untouched. It is today as nearly all of Europe once was, and it's the closest thing we have to the forest that once blanketed the eastern half of the United States. In the Białowieża, oak trees grow to be over 20 feet in circumference. Wisent, the European bison, graze in dappled meadows much as American bison once did. The space between the great trees is damp, thick with centuries of leaves and discarded branches. Mushrooms and other fungi grow in riotous profusion. Few green plants exist below the canopy that is dense enough to keep the world below in permanent gloom. Here in the old forest, believing in monstrous beasts and sacred groves is easy. This is the forest not just of our ancient ancestors, but of those who came only a few generations before us. This is what temperate land does when we don't get in the way. What we have today is a shadow. A ghost forest.

If the Białowieża seems like a long way to travel to see what America's forests were like centuries ago, at least there is a Białowieża. You can still seek out the Cumberland Road as well. US Highway 40 follows much of the original route between Illinois and points east. However, to see an example of an economy based on the American School created under George Washington and shaped by subsequent administrations, you'd have to travel to . . . well, nowhere really. Certainly nowhere in America.

For more than a century, the American School of economics would be based on three ideas: supporting industry through tariffs and subsidies in opposition to free trade, government creation of improvements to help commerce and investment and control of private infrastructure, and a government-controlled fiscal infrastructure that would encourage growth of the economy through regulation of credit and direct intervention in banking. During the administration of Abraham Lincoln, two more ideas were added to the canon:

- Support for public education and research grants as a means of advancing science
- Ensuring that government policy did not drive "class struggle" by further enriching factory owners at a cost to workers

Lincoln's economic advisor, Henry Carey, compared the American system with the British system of more laissez-faire economics in his book, *The Harmony of Interests:*

> Two systems are before the world. . . . One looks to increasing the necessity of commerce; the other to increasing the power to maintain it. One looks to underworking the Hindoo, and sinking the rest of the world to his level; the other to raising the standard of man throughout the world to our level. One looks to pauperism, ignorance, depopulation, and barbarism; the other to increasing wealth, comfort, intelligence, combination of action, and civilization. One looks towards universal war; the other towards universal peace. One is the English system; the other we may be proud to call the American system, for it is the only one ever devised the tendency of which was that of elevating while equalizing the condition of man throughout the world.[3]

Not until 1893 did Democrat Grover Cleveland begin reducing the tariffs that had marked the American system. The reduction was the start of a general disengagement between government and the economy—a surrender to the British system.

These days even an enormous crisis in the economy is not enough to make us consider a return to the American system. Protective tariffs, strong investment in education and jobs, standing up for the workers and against corporations all seem a bit . . . *quaint*. More than one voice would be quick to declare these ideas (the ideas of Washington, Hamilton, and Lincoln) to be socialist. Even un-American.

A few scattered groves of American Chestnuts still exist, and multiple efforts are under way to restore some semblance of the old forest to at least some parts of the nation. But American economics may be truly extinct.

American Wildcat

JOHN JAMES AUDUBON

CHAPTER 12

Little Giants

Hey, let's be careful out there.

SERGEANT PHIL ESTERHAUS (1981)

Between 5000 and 7000 tigers live in the United States. That there should be that much uncertainty about just how many 500-pound cats are sharing our nation is a little surprising, since even the difference between those two numbers—the 2000 tigers that we *may* have—is more than the entire population of tigers remaining in India.

All of the American tigers are supposedly behind bars. Most are in zoos, but a large number are involved in some type of entertainment (circuses, stage acts, and so on). An insane number (in this case, *insane* being defined as any number greater than zero) are in private homes as pets—a practice still allowed by over a dozen states. Two tiger sanctuaries, where a score of aging tigers slumber, are located only a few miles from my home. One of them is a combination tiger sanctuary and golf course, which would seem to offer some unique opportunities for "hazards."

Across the country, there are frequent reports of tigers or other large exotic cats spotted in the wild—if slouching through the shrubbery at the edge of a subdivision can be considered wild. No doubt a few tigers and other exotic animals have slipped loose over the years. Still, even if all the doors on all those tigers were to slip open at the same time, it would certainly result in temporary chaos and more than a few . . . incidents. But in the long term, the results are easy to predict: all the tigers would die.

To see why, you have only to look at the tiger's relatives in the Americas. A male mountain lion weighs in at about 150 pounds—about a quarter of the size of a large tiger. To keep itself fed, a mountain lion prowls an area ranging from around 50 to as much as 400 square miles. The American bobcat is a considerably smaller animal. At 25 pounds, it's about one-sixth the size of the mountain lion, and it makes do with about one-sixth the area. In the eastern United States, only Florida still houses a tiny population of mountain lions (barring the few who may have been turned out by exotic pet owners having second thoughts), but bobcats number in the hundreds of thousands.

Just as it's hard for large animals to make it through the narrow gates of extinction events, it's also hard for them to survive when big areas are broken into smaller and smaller lots. Supporting a large predator, like the mountain lion, requires a large area in part because a large predator needs a large amount of prey. Warm-blooded mammalian hunters down a goodly amount of meat each day. (For an adult mountain lion, that amount is around 10 pounds.) Mountain lions will eat anything down to mice and insects, but supporting a daily meat habit that runs through the equivalent of a freezer full of steak requires a lot of meat on the hoof (or on the L. L. Bean hiking boot). Squeezed into little islands of territory between unfriendly people—who hunted them for bounties—mountain lions didn't make it.

Little bobcats made it just fine, thank you. In fact, bobcat numbers today may be greater than they were a century ago when they had to compete for prey with their larger cousins. When territory gets small, the small get successful.

To support large animals one needs large areas. That's true even of animals that dine on plants. Big plant eaters have serious advantages over smaller animals when it comes to exploiting their environment. Elephants can push over trees to get at the leaves or uproot shrubs to chew the roots—options not open to their smaller competitors. This solution is great for the short term, but an animal that survives by shoving over trees won't last long if only a few trees are available. Elephants can make good use of their size, as long as they have an equally large area to use.

Elephants' ability to exploit a wide variety of resources makes elephants viable in a surprisingly wide range of environments. Yes, they can be found trekking across the African savannas, but they are also creatures of the deepest forests and are even found in vast deserts. The elephant's shaggier relatives of 12,000 years ago must have accrued similar advantages. Mammoths and mastodons were able to survive in tough conditions, both because their size helped insulate them against cold temperatures and because they were able to go after everything from roots to tree-tops in an effort to find food.

Of course, elephants and all their kin are still relatively small when compared to the real giants of the past. The largest elephant on record weighed in at 13 tons and stood a bit over 14 feet at the shoulder. The sauropod dinosaurs of the Jurassic and Cretaceous periods included animals that weighed as much as six elephants and some whose length was well over 120 feet.

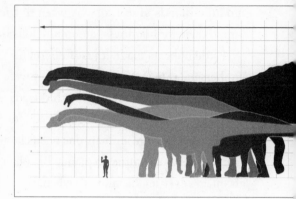

FIGURE 12.1. SIZE
COMPARISON OF
LARGEST SAUROPODS

MATT MARTYNIUK, 2007

Like elephants, the sauropods were surely able to exploit food resources that other animals couldn't have reached. Restorations have portrayed them as anything from lumbering vacuum cleaners hoovering up ferns at ground level to hyper giraffes browsing among the redwoods, inspiring debates about just how high sauropods could raise their long necks. The truth likely falls in between. At least some sauropod necks provided enough flexibility to grab food from ground level to 40 feet up in the branches or higher (see Figure 12.1).

When these giants were first discovered, it was assumed they were too heavy to stand without the assistance of water. The earliest reconstructions we have of such dinosaurs have them standing in swamps, munching on soft water plants—the image that persists for many people. But further study shows that these dinosaurs were fully capable of walking quite well across dry land. In fact, they were likely creatures of the open plains, marching many miles over prairies of ferns and mosses in search of pastureland. The evidence indicates that the areas in which many sauropod bones are found were the Mesozoic equivalents of savanna. These animals may well have also taken advantage of the

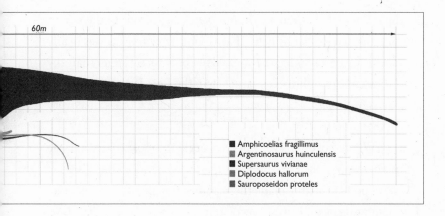

60m

■ Amphicoelias fragillimus
■ Argentinosaurus huinculensis
■ Supersaurus vivianae
■ Diplodocus hallorum
■ Sauroposeidon proteles

food available around water—there are several known trackways along ancient shorelines. Like elephants, they may have also been able to cross deserts and reach resources beyond the ability of smaller creatures. However, one area where elephants currently thrive may have been out of reach for the sauropod giants—it's hard to imagine such animals living in the forest, if only because turning around a 100-foot form among tightly clustered trees seems impossible. On the other hand, they might have adopted a version of the elephant's strategy: knock the trees down first, then turn around wherever you want.

The largest known species of sauropod is also the most mysterious. In 1877, American paleontologist Edward Drinker Cope pulled two vertebrae and a single leg bone from a quarry in the Morrison Formation of Colorado. The bones from this quarry tended to crumble once exposed to air, and Cope lacked the preservatives available to later paleontologists. This particular set of bones was especially delicate, and Cope named the creature they came from *Amphicoelias* ("doubly hollow") because the remains seemed to be as much air as bone. The bones he found have since disappeared; it may be that Cope simply discarded the broken bits. In the meantime, he drew a sketch of just one of the bones. That sketch led to estimates that a living *Amphicoelias* was nearly 200 feet long—making this animal not only vastly larger than any other sauropod, but even larger than the Blue Whale. No specimens of *Amphicoelias* have been found since Cope's discovery in 1877.

When sauropods were at their peak, the most common large plants were conifers—pine, cedars, yews, and so on. The big sauropods probably made these plants a staple of their diet. Not many animals dine primarily on pine needles today because these prickly little leaves are notoriously low in nutrition. That's where another advantage of size comes in. Big animals pack around a big gut, and big guts on average are more efficient at extracting nutrition from food. As sauropods expanded their diet to include less and less nutritious sources, they needed to pack around bigger and bigger stomachs—a kind of arms race on limited calories and vitamins as these creatures attempted to nab every source of food possible.

Giants of other sorts also require large habitats, like the vast audiences needed to feed giant television shows.

Steven Johnson's 2006 book, *Everything Bad Is Good for You*, makes a convincing argument that over the years television programs have increased in complexity and quality. While a nostalgic fan of television past might groan at the idea, a few minutes of watching the best shows on current television—versus what was on 20, 30, or more years ago—is enough to conclude that Johnson is right. Would anyone really argue that *Mannix* is a more worthy program than *Mad Men*? Would you rather spend your time in Hawaii with *Five-O* or *Lost*?

Starting in the early 1980s, television hit a period of rapid development, and the hour-long drama in particular grew into

rich, complex entertainment. The niche once occupied by vacu-
ous, slow-moving dramas (built around single plot threads with
minimal character development) grew increasingly crowded with
programs featuring expansive casts of experienced actors, complex interwoven multi-threaded plots, and intricate story arcs.
Problems were no longer neatly buttoned up in the space of a
single episode, and characters were no longer neatly brushed off
and restored to their starting positions for the beginning of the
next week.

Many of the conventions that helped primetime "quality television" reach that apex of respect came from the least-respected
portion of the daytime airwaves—soap operas. In soap operas
television worked through the problems of dealing with serial
material, large casts, and ongoing stories. The nighttime dramas
benefited from larger budgets, which in turn allowed hiring of
more experienced actors, writers, directors, and staff, along with
more lavish sets and effects. But complex primetime shows were
night-dwelling descendents of their soapier kin.

Many of the soap operas had made the transition from radio
incarnations. They also drew inspiration from the "serials" once
shown before the main feature at movie theaters. Ultimately both
soap operas and nighttime dramas owe a debt to the kind of
serialized fiction pioneered by writers such as Charles Dickens,
who published his works a few chapters at a time in magazines.
Desperate readers, who haunted the docks in New York City in
1840 waiting to hear the fate of Little Nell in *The Old Curiosity
Shop*, are the spiritual ancestors of viewers who browsed the *TV
Guide*, hoping for a clue about whether Uncle Junior or Tony
would emerge as the head of the family.

Unfortunately, there was a problem in television land. Just as
large dinosaurs had a hard time making it in shrinking habitats,

big television was facing its own dilemma. And just as it did with the dinosaurs, the threat originated in space.

Johnson isn't alone in singling out *Hill Street Blues* as the definitive punctuation in the television equilibrium. It was the program that proved that the viewing audience had grown up, that we had mastered the conventions of television well enough to deal with multiple overlapping plot lines and characters whose motivations were considerably less clear than those of Marshall Matt Dillon or Sheriff Andy Taylor. The world in which *Hill Street* first clambered over the television plains was one of only three networks. Even so, the show struggled to gain a footing, never making the top 20 shows in its seven seasons. Still, *Hill Street* set the pattern that would be followed by a hundred dramasaurs to follow.

Then came cable. And satellite. New broadcasts networks came (and went). A viewing audience that had been divided into thirds further split, first faced with dozens, then hundreds of choices. Feeding an hour-long dramasaur became increasingly challenging as the environment was reduced to a series of smaller and smaller cable islands. The same economics is at work in the daytime, reducing the once thundering herd of soap operas to a bare handful of gaunt survivors. Increasingly, reality shows—descendants of *Arthur Godfrey's Talent Scouts*, which was television's top show in the first full year of Nielsen ratings—shoved aside the ungainly giants.

Even the most gaudy reality shows cost a fraction of what it takes to film a scripted drama. That difference paid off in television's early days when only a few people owned sets, and the budget advantage was increasingly telling as revenues shrank. In 2009, NBC threw in the towel on supporting a full slate of dramas and replaced most of its evening tally with five nights a week of Jay Leno. All across the digital dial, giants were looking fragile.

There is an alternative. In the Late Jurassic rocks of northern Germany are the remains of a sauropod named *Europasaurus*. Its relatives among the brachiosaurs weighed as much as 90 tons and were over 100 feet long, but *Europasaurus's* giant ancestors were stranded on an island in the shallow sea that covered many parts of Europe at the time. There they underwent a process that's been seen many times over evolutionary history but which is still not fully understood—island dwarfism. In this reduced environment, "fitness" no longer favored the biggest animals among the sauropod herd. Instead it was the smaller members who were able to find enough food in the limited space. Over generations, the *Europasaurus* was reduced from a Jurassic Park giant to a size closer to the Flintstone version of Dino. An adult was around 15 feet long and held its head at a level that would put it eye to eye with a human.

Europasaurus wasn't the only mini-sauropod. Several others underwent the same process of shrinkage. *Magyarosaurus* lived on islands in what is now Romania and was even smaller. These animals were still clearly sauropods with the same design as their big relatives. Just smaller.

Island dwarfism isn't limited to dinosaurs. Skeletons of tiny *Homo floresiensis* found on the Indonesian island of Flores show that humans can be subject to the same phenomenon. Members of this human offshoot—which survived long enough that the last of the "real hobbits" must have shared the island with modern humans—were only half the size of their mainland neighbors. Adults stood only a bit over three feet tall.

On three of California's Channel Islands are remains of another little giant. Mammoths reached the islands around 50,000 years ago and proceeded to follow the isolated-island path of reduced size. An adult Channel Islands Mammoth was around five feet tall at the shoulder—the Shetland pony of the mammoth family (and in fact, Shetland ponies are also examples of island dwarfism). The little mammoths continued to browse on the Channel Islands until ancestors of the Chumash people showed up around 11,000 years ago and were less impressed with the mammoths' cuteness than they were with their tastiness. Even so, the Channel Islands Mammoths far outlived those a few miles away on the mainland.

Dramasaurs may survive in the same way. The average episode of *Mad Men* costs about $1.1 million. Not exactly cheap, but about one-fourth the cost of a typical network hour. Of course, the average audience for *Mad Men* is about one-thirtieth that of an episode of *Dallas* at its peak, but newer shows have also extended their search for viewers into the DVD aisle and into complex cross-channel promotions. In addition to shrinking their per episode cost, dramas living in the islands of cable have reduced the number of episodes they shoot in a season.

This doesn't mean the quality of the shows has decreased. In fact, the pressure to survive in this environment may be driving these shows toward even more innovation, riskier performances, and more challenging plot lines. They are still giants—just smaller giants.

After all, in natural selection, *small* doesn't mean less fit. It often means successful. Tigers in the wild would never survive in America's heavily divided ecosystem. Even the native mountain lions have barely been present over most of the country for a century. But the bobcat carries on.

Not only do small creatures stand a much better chance of surviving periods of widespread extinction, they can also outlive their bigger relatives. On Wrangel Island, a hundred miles off the coast of Russia in the frigid waters of the Arctic Ocean, the last

Woolly Mammoths on Earth outlasted the giants of the plains by almost 10,000 years. Their relatives might have been impressive, but they were long gone. Meanwhile, Wrangel's little mammoths outlived 14 Egyptian dynasties, the Kingdom of Hammurabi, and the Xia dynasty in China.

Giants might be exciting, but little giants are built to last.

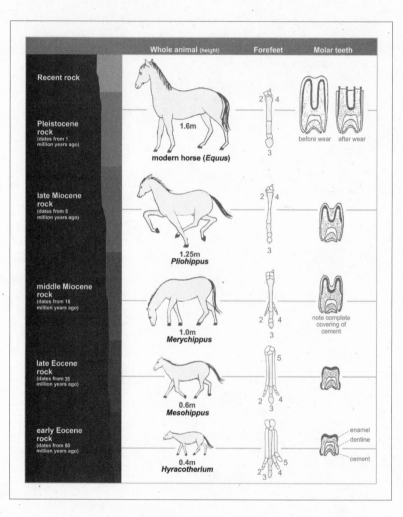

Evolution of the Horse

ALEX BROLLO, 2006

CHAPTER 13

Mustangus americanus

The best white man I've ever seen.

RED CLOUD,
discussing O. C. Marsh (1875)

In 1874, Othniel Charles Marsh published a paper on the toes of horses—a paper that is among the most widely reproduced, referenced, and adapted scientific findings in history. It may seem strange that a paper on a subject so esoteric should have such an impact, but timing is everything.

Charles Darwin's *On the Origin of Species* was published in 1859 and quickly became the focus of praise and controversy on both sides of the Atlantic. In his work, Darwin used a variety of living creatures as evidence in support of his theory of natural selection but made only light use of fossil forms for several reasons. First, Darwin wasn't familiar with many of the most ancient forms of extinct creatures and wasn't comfortable in making connections between those creatures and the ones still living. Darwin also admitted that the fossil record was sparse and contained many gaps.

Critics immediately seized on this point, demanding that those favoring natural selection produce specimens that provided convincing support for Darwin's theory. Without the backing of the fossil record, Darwin's opponents declared that natural selection was little better than informed speculation. In particular, critics pointed at the lack of transition specimens in the fossil record. If creatures had undergone such astounding alterations over time, as Darwin's theory required, then why were the stones so full of things that seemed to be all this or all that?

Yes, the sandstones might cough up fish not like those found dangling from a fisherman's hook, but they were still fish. Reptiles of Unusual Size might have been squeezed between plates of stone formed from primordial ooze, but they were still reptiles. Where were the specimens caught in mid-step between one thing and another? Where were the fish with the legs of amphibians? The reptiles with the teeth of mammals? Where was the proof written in the rocks?

As it happened, a remarkable specimen did appear only a year after *On the Origin of Species* reached British shelves. The specimen was turned up by workers at a limestone quarry in Solnhofen, Germany. The limestone there had been formed in the still waters of a lagoon at a time when much of Germany was an archipelago on the edge of the vast Tethys Sea. The waters of that vanished lagoon were so calm that the deposits there were as fine and uniform as layers of dust drifting down in an abandoned room. The limestone produced from those sediments was particularly well suited for carving plates used in lithography, a type of printing that was then very popular for reproducing images.

The fine-grained nature of those stones also meant that the remains of plants and animals that fell into that ancient lagoon were preserved with startling fidelity. Most fossils preserve only the hard parts of the original creatures—bones, shells, and teeth. But at Solnhofen delicate features could still be seen clearly: the tiny tube-feet of starfish, the veins of a dragonfly's wing, or—as in the case of that specimen in 1860—the structure of a single feather.

The feather pre-dated the remains of any bird known at that time by several million years. The source of the lonely feather was the center of heated debate, but the debate didn't last very long. In 1861, the same quarries produced another specimen, one that would eventually come to be called *Archaeopteryx lithographica* (Figure 13.1).

Here was an animal that had the feathers and wings of a bird, but the clawed hands and long bony tail of a small dinosaur.

That first specimen was headless, and for a time those scientists who had a vested interest in denying evidence in support of Darwin's theory suggested that the claws and tail represented only a very odd bird. This insinuation set off a race to find a fossil bird that was more complete—a race that was won when O. C. Marsh located fossils of a Cretaceous bird he named *Hesperornis* in the chalk of Kansas. The stunted wings on the remains of *Hesperornis* showed that it was flightless—it was more like a six-foot-tall penguin than a condor. But this penguin had teeth. The remains of Marsh's giant bird, along with more complete specimens of *Archaeopteryx*, forced the doubters to face the rest of the puzzle. These ancient birds not only had the tails and claws of reptiles, they also had heavy skulls and teeth rather than beaks. They were true transition fossils between dinosaurs and modern birds.

The detail recorded in the best specimens of *Archaeopteryx* is staggeringly sharp. These specimens are not a handful of bones or teeth from which scientists, using logic and cladistics, have

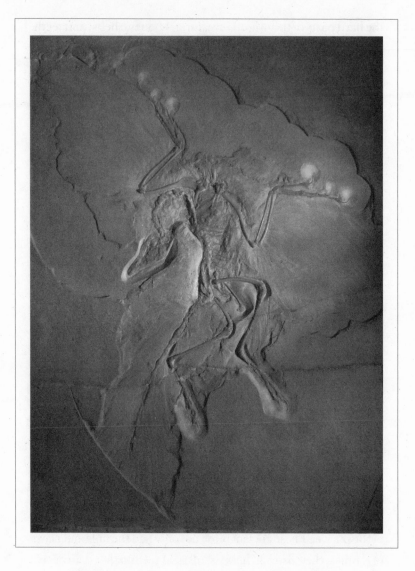

FIGURE 13.1. *ARCHAEOPTERYX LITHOGRAPHICA*

molded a whole theoretical beast. They *are* the beasts. You can see the structure of each tiny tooth. You can study the intricate rachis and barbs of each feather. If Darwin had advertised for proof for his new theory, he could scarcely have imagined a better pitchman than the tiny dinosaur-bird from Bavaria.

Still, a few critics of natural selection averted their eyes and refused to look on the new specimen as substantial proof of Darwin's theory. *Archaeopteryx,* they said, was merely some sort of sport. The specimens represented some dead-end mutation based on a bird that happened to mimic some characteristics of reptiles or perhaps a mutant reptile that sported particularly feather-like scales. Besides, it was only a single step on the trail between reptiles and birds. If *Archaeopteryx* was halfway between the birds and reptiles, where was the creature that was one-quarter reptile, three-quarters bird? Where was the three-quarters bird, one-quarter reptile? *Archaeopteryx* was (then, at least) a solo act. Where was a complete sequence of creatures shading from one type thing to another?

And that's where the horses come in.

O. C. Marsh was not always an easy man to love. Yes, he could be quiet, introverted, and thoughtful. But he could also be distrustful and quarrelsome. By the 1880s, he would nearly bankrupt himself (and find himself socially marginalized) because of his long-running feud with fellow paleontologist Edward Drinker Cope.

But the "bone wars" came later, and Cope bore the lion's share of the blame. The horses came first.

Marsh really had three things going for him. First, he was bright—really, stunningly, bright. He made intuitive leaps that eluded many of the great minds of the day. He had a massive grasp of morphology and a nearly boundless imagination—and he put them together in a way that few then (or now) could match. Second, he was a workaholic. Marsh was rarely happy unless he was turning a bone over in his hand or dashing off one of his hundreds of articles. He named new taxa with the regularity of a scientific machine gun. Third, he was the nephew of a famously wealthy philanthropist, which is very rarely a bad thing for someone who wants to spend his time in a pursuit not particularly known for its commercial potential.

At Marsh's urging, his uncle George Peabody founded the Peabody Museum of Natural History at Yale, ensuring that Yale would be (and still is) a center of paleontological research. Not so coincidentally, Marsh was named to America's first professorship in paleontology that same year. In a world full of amateurs, Marsh was a professional. He had brains, drive, a steady paycheck, and dozens of willing Yale undergrads ready to lug his bones from the hills and badlands of the West. Before Marsh, 9 species of dinosaurs were known. In the next two decades, Marsh's teams found 80 more. (His rival Cope identified 56 species of his own.)

By 1874, Marsh had been collecting fossils of horses for several years. He was far from the first: a book about American horses had been published as early as 1848. Charles Darwin himself had found the tooth of a fossil horse in Argentina back in 1833 while taking part in the *Beagle* expedition. William Clark (the Lewis and Clark, Clark) found fossil horse bones in Kentucky in 1807 and shipped a few to his friend Thomas Jefferson.

What made all these bones so significant at first was the fact that there were no horses in America. At least, there had not been any horses when the Spanish arrived at the end of the fifteenth century. Neither Clark (nor Jefferson, who had an interest in paleontology) seemed to notice the import of that fact, but Darwin certainly did. The fossils showed that there had been horses in the Americas, but those horses had disappeared. It was further proof that the collection of animals roaming the continents was not a constant. This was even more true when it turned out that Darwin's tooth came not from a modern horse, but from a variety no longer found anywhere in the world.

By Marsh's time, a number of different types of fossil horses had been collected in the Americas. What Marsh did next seems simple—he put them in order. But although this may sound no more amazing than arranging the ponies on a carousel, one look at the creatures in Marsh's chart would show that it wasn't quite that simple.

At the start of the line was a creature called *Hyracotherium*. (Marsh knew it as *Eohippus*.) The scientific name alone is enough to suggest how different this critter was from a modern horse. The name means "hyrax-like beast." Like a modern hyrax (rabbitish animals found in Africa and the Middle East), *Hyracotherium* was small, about two feet long. If you were to see one of these animals crossing a field, "horse" is not exactly the first idea that would cross your mind.

In addition to the size, there was one other very distinct difference between this 50-million-year-old creature and a modern horse. *Hyracotherium* had four toes on each foot. The modern horse, *Equus*, has a single toe ending in a thick hoof. Hidden, internal remnants of two other toes are still present in a modern horse, but they've been reduced to tiny spurs of bone located high up on the leg. Each step in Marsh's chart of horse ancestors follows the same trend—reduction of toes, increase in size, lengthening of the head, and more complexity of teeth. First four toes, then three toes, then a gradual reduction of the other toes as the central toe becomes capped by the hoof we see today. These feet were mounted on animals the size of dogs, then deer, then ponies, and finally the imposing modern horse. The teeth move from small molars with nearly flat surfaces to larger, more complex teeth set in long, increasingly *horsey* skulls. All these steps tell the same neat story. Here are animals making the transition from small forest-dwelling browsers to large wanderers of the open plains that graze on tough grass, animals that depend on speed and strength to deal with predators.

Marsh had produced not a single transitional specimen, but a whole sequence by which species moved through drastic change. Without all the intermediate stages, even an anatomist as skilled as Marsh might not have recognized the tiny *Hyracotherium* as even being an ancestor of the modern horse. But by placing related species on a timeline, Marsh reduced drastic change into a series of easily understood steps. It was exactly the sort of succession of creatures that you might expect to see resulting from natural selection. Best of all, it was essentially correct—where it wasn't wholly misleading.

The sequence that Marsh produced does reveal major points along the path that horses took in going from tiny forest animal to lord of the plains, but that path wasn't taken across a blank sheet of paper. The species that appear in Marsh's chart are only a small fraction of the ancient horses that once existed—just a sampling of a variety more vast and wonderful than any A-to-B-to-C chart can reveal. Some ancient horses didn't lose those extra toes. Instead, their toes grew long as the feet adapted to walk over softer ground. Some developed teeth better suited to nipping leaves off trees than grazing on grass. During the Miocene epoch (from about 23 to about 5 million years ago) more than a dozen species of horses roamed the Americas. Rather than a simple line marked by a few species, the horse family was a large, complex "bush" producing diverse creatures, only a few of which looked particularly horsey.

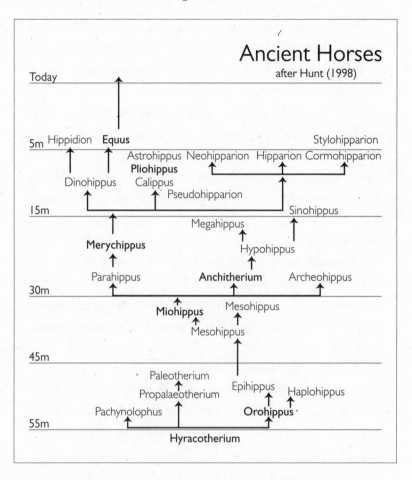

FIGURE 13.2. CHART OF HORSE EVOLUTION

A better look at the evolution of the modern horse would be something like the chart in Figure 13.2. Those species in Marsh's straight-line evolution are shown in bold. And this is still only a fraction of the diversity that the horse family displayed during the last 50 million years. Although all the animals selected by Marsh were indeed relatives of the modern horse, they were not all in the direct path that led from *Hyracotherium* to Seabiscuit.

What makes it tempting to draw a simple picture of the family tree of horses is the simplicity of the situation we see around us. Starting around 4 million years ago, as the world grew cooler and drier, the horse family tree started to . . . thin out. Hipparions, very successful and widespread grass-eaters similar in size and build to a modern horse, gradually disappeared from Asia, Europe, Africa, and North America. As ice expanded and contracted around the northern hemisphere, the non-*Equus* horses disappeared and even the members of that surviving genus were greatly reduced. The last of the near-horses held on until around 8000 years ago. That's also when *Equus scotti* and all the other North American horses left the stage. For the next 6500 years, there would be no horses in America.

Four and a half centuries after the Spanish brought horses back to America, another sort of pony was born in Detroit, Michigan.

In 1964, Ford division general manager Lee Iacocca recognized an open space in the automotive ecosystem. At the time, Ford had successful large sedans, a popular pickup, and a newly introduced compact car, but none of these vehicles was particularly sporty. The Thunderbird had been released in 1955, intended as competition for the Chevrolet Corvette. But despite outselling the Corvette by 23 to 1, Ford executives (in particular the soon-to-be secretary of defense, Robert McNamara) worried that the two-seater roadster had a limited market. By 1964 the Thunderbird had two more seats, a thousand pounds more bulk, and an upscale price tag as a "personal luxury vehicle." It was a sales success, but it wasn't exactly a sports car.

Ford introduced a new vehicle. That vehicle didn't pop up out of nowhere. Just as the bones of *Equus* reveal a relationship to earlier creatures, the Ford Mustang's ancestors were not hard to see. The chassis, suspension, and drive train components came from parts used in the Ford Falcon and the Fairlane. Like the early birds, whose design wasn't up to prolonged flight, this first Mustang still carried the baggage of its ancestry. It wasn't very fast. It wasn't very sporty. It really wasn't much of a radical departure from other Fords already on the market. However, it was attractive and it was cheap.

Competitors dismissed the little Ford as a "secretary's car," but there were a lot of secretaries—and a lot of non-secretaries who liked the idea of a fun, inexpensive auto. In the world guided by a different form of selection—consumer choice—the Mustang fit a big, empty niche. Ford's marketing department had forecast sales of 100,000 units during the Mustang's first year, but the factories were soon working overtime as the Mustang flew off the lots. A million were sold in the first 18 months.

As with the ancient horses of the Americas, it's tempting to group the Mustang into a neat sequence of models. After the original model came the Mustang II (which appeared very late in the Preoilcrisian epoch of the Interstatian Age), then the "Fox" platform of the 1980s, followed by the edgier design of the 1990s and the slightly "retro" model in showrooms since 2005. But, just as with the flesh and blood mustangs, the picture isn't that simple.

Ford's Mustang showed an ability to kick off new "species" and "subspecies" that would be the envy of the natural world. By 1968, there were nine different base models of Mustang. Coupes, convertibles, and fastbacks were manufactured with a bewildering array of engines, transmissions, and extras. Over a period of just three years, 13 different engines were available. And that's not even counting the many semi-official models being cranked out by racing firms that modified the Mustang into more power-ful machines.

Something else began happening to Ford Mustangs, some-thing that had also happened to those other mustangs: over time they got bigger and heavier. Each revision of the car from 1965 to 1973 brought on a meatier and more expensive vehicle. This kind of growth is a path common to both animals and cars. The Honda Civics of the natural and automotive world move from tiny econo-boxes to more substantial forms, leaving room for the next round of little Fits to take their place. The strategy is not particularly poor—so long as your ecology stays healthy and food is plentiful.

As the Mustang was growing larger, it was also growing competitors, and its success brought on a whole generation of "pony cars." Most of these were larger and more powerful than the original Mustang as the manufacturers fought to win the con-sumer selection derby through providing more room and power. By 1973, Camaros, Javelins, Firebirds, and Cougars were all scrambling for the same "ecological niche" as the Mustang—just as many other forms of horse (and other grazers) had challenged *Equus* on the plains. And just as ecological challenges starting in the Pliocene whittled away at the horse ranks, a disaster was on the way that would soon reshape the pony car competition.

On October 17, 1973, OAPEC (the Organization of Arab Petroleum Exporting Countries) announced an embargo against Western countries. The price of oil tripled almost overnight. Within a few years, it would triple again. For the pony cars—which fed off cheap gas and the perception that powerful auto-mobiles were an expression of freedom—the years of easy growth had ended. The verdant fields of buyers dried up. An automotive ice age had arrived.

Many of these models disappeared. Some of the nameplates would show up on new vehicles and some were gone for good. But Mustangs lived on. In fact, the Ford Mustang has been in continuous production since 1964.

What made the Mustang the automotive equivalent of *Equus*? Why was it left alone, the only surviving genus out of a once-complex and diverse field? It was a combination of good planning and good luck.

Oil prices had been rising for several years before the 1973 embargo—not sharply, but enough so that Iacocca and other executives at Ford were starting to be concerned about the Mustang's growing size, power, and thirst. In addition, they were being nudged from another direction. Fans of the original 1965 Mustang wrote to Ford complaining that by the early 1970s the car had lost the cheap, fun appeal of the original. It had become a boy racer's car rather than a secretary's car, and out there in the ecosystem of auto buyers, boy racers were just not that common. Ford went to work, trimming down the Mustang and designing a new car based on the Pinto subcompact. That new car, the Mustang II, premiered just weeks before the oil embargo. The new car was smaller, it was cheaper, and it was designed to be far more reliable. It also offered a fit and finish that was well ahead of most American cars at any price.

Dismissing the Mustang II has become fashionable these days. In fact, it's often treated as a model best forgotten, not a "real Mustang" at all. With a four-cylinder engine in the base model, the new Mustang was spurned by the audience that had come to love its larger and more powerful predecessor. But the Mustang II got much better gas mileage than its competitors. The new car's smaller size and reduced power made it a loser in a straight-line duel with anything other than the pokiest economy car, but its reliability and good value for the money positioned it to compete with the flood of import cars that moved into the market as oil prices rose. Those who dismiss the car today forget that in 1974, the Mustang II—the multiple-award-winning, Car of the Year, Mustang II that outsold the previous model in the face of an auto market as shaky as any seen until 2009—was one of Detroit's few bright spots.

The Mustang survived. And by squeezing through that tight spot, it went on to spawn new variants and expand into new niches. It got the secretaries *and* the boy racers (and the girl racers, as one of my friends, who recently bought one of the Mustang's more potent subspecies, will attest).

Both the Ford Mustang and the mustang mustang are American originals. They have spread around the world, but they started here. Both started from a simple beginning, spawned dozens of variations, and were pared back by calamities. Both have proven tough enough to take on a changing world.

Finally, both are also examples of one of those great rules of selection—luck counts. *Equus* didn't lift a hoof to the wind, sense the coming ice age, and decide it was time to leave the forest for the plains. It was already well adapted for the world that was coming. Ford didn't launch its drive for a post–cheap oil Mustang in the wake of the crisis. The new car was already on the market when the gas lines appeared. Timing may not be everything, but whether it's the success in the marketplace or plain old survival, it means a lot. With few exceptions, species don't thrive by adapting to changing conditions. They survive by already possessing some quality that makes them pre-adapted for what's coming.

As we look ahead and see challenges on the horizon—including the looming threat of climate change—we have to remember that the survivors are not those species that learn to swim after the flood arrives. You need to be lucky, or smart, enough to take steps in advance. Otherwise you'll be a *Hipparion* or, worse still, an AMC Javelin.

Charles Darwin

ELLIOT & FRY, 1881

CHAPTER 14

Galluping Science

There are three kinds of lies:
lies, damned lies, and statistics.
BENJAMIN DISRAELI

On the bicentennial of the birth of Charles Darwin in 2009, the Gallup organization published the results of a poll that asked American adults whether they believed in the theory of evolution. The headline of the article Gallup released to news organizations around the world was quite dramatic.

ON DARWIN'S BIRTHDAY, ONLY 4 IN 10
BELIEVE IN EVOLUTION[1]

The pattern they set was matched elsewhere. Whether it was the AP, CNN, or *USA Today*, the results of this poll were headlined everywhere with the word *only*. Only 4 in 10. The exception was Fox News. They went with "*Fewer* Than 4 in 10 Believe in Evolution."

Over the following months, the poll was cited as proof of America's spiritual nature, as proof of America's lack of scientific education, as proof of America's conservative values, and as proof of America's plain old ignorance. The poll was used by religious groups to argue against the teaching of evolution in schools. It was used by educational groups to argue for more science education in schools. Only a month after its release, the Texas Board of Education cited the poll as they split over requirements for science education. The board did not go so far as to require that the teaching of evolution be banned from the classroom—as the chairman of the board wanted—and it did not order that teachers must advocate creationism instead. It presented "compromise" language directing the science teachers to be sure that instruction included "examining all sides of scientific evidence of those explanations so as to encourage critical thinking by the students."[2] This language was enough to satisfy those seeking a crack through which they could insert creationism into the classroom.

A year later, the results of the Gallup Poll were still being cited, this time as a film company attempted to find distribution for a film about Darwin's life. American distributors shied away, saying that the poll showed evolution was "too controversial" for audiences in the United States.

However, the data in that very same survey indicate that American attitudes toward evolution are far different from what the headline suggests. Forget for a moment your own personal position on evolution. Put aside what you know of Darwin's supporters and opponents. Think about this question on its own: is evolution controversial in the United States?

To help you decide, consider this headline from another national poll, this one conducted by the firm Research 2000.

ONLY 4 IN 10 AMERICANS BELIEVE IN
THE THEORY OF CONTINENTAL DRIFT[3]

Continental drift, a theory developed in the early twentieth century by German meteorologist Alfred Wegener, is not just accepted by every practicing geologist, it's directly measurable. Sensors placed on the crust of Earth can detect the motion of each of the continents and the plates on which they ride. We can see that the African Plate is currently heading northeast at a bit over two centimeters (one inch) per year. The North American Plate is moving southwest at around half that speed, while the Pacific Plate—including areas along the West Coast of the United States, are galloping along four times as quickly. That difference in motion between the North American and Pacific Plates helps to explain why so much of the West Coast is a geologic fun zone.

Backtracking the motion of the continents, we can see that around 250 million years ago the world's significant land masses were gathered into a "super continent" that geologists call Pangaea. This was only one of several such get-togethers held over the past few billion years, and 300 million years from now the continents will all be grouped together again. Not only can we see the motion of the continents today, but the past activities are confirmed by the location of fossils, by deposits of rock shared along continental boundaries, and by magnetic fields frozen into the wandering stones.

Except for that very small percentage of people who insist that the world was created *ex nihilo* in a matter of days, continental drift seems as if it should be a fairly innocuous idea. So how could any national poll find that so few Americans believe in something so completely and extensively documented, so thoroughly established, and so absolutely tangible? Should we worry that America is hopelessly backward and mired in some kind of anti-science Dark Age?

Not quite. The question that was asked in that Research 2000 poll that generated those results is in Chart 14.1.

CHART 14.1. Research 2000 Poll on Continental Drift

Question: Do you believe that America and Africa were once part of the same continent?

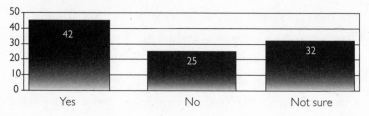

What does this show? Well, it shows enough to generate a misleading headline, but not much else. Both the question itself and the way the results of the survey are presented are so deeply flawed as to be worthless. They are, in fact, quite intentionally idiotic. (And in case you're worried that it wasn't intentional, I can assure you that it was—I wrote the question.)

The question was written to press emotional hot buttons. It wasn't created in dry language concerning the current motion of the plates. It's not "Were all the continents once merged into Pangaea?" or "Was the North American continent once closer to Europe?" or simply "Do you believe in continental drift?" The question was framed with the intention of reaching down and touching something more than just the understanding people have of a particular area of science. It's "Were *America* and *Africa* once touching?"

If you think that doesn't matter, I invite you to look at the regional breakdown of results (Table 14.1).

TABLE 14.1. Research 2000 Poll
on Continental Drift, Regional Breakdown

Do you believe that America and Africa were once part of the same continent?

REGION	YES	NO	NOT SURE
Northeast	50	18	32
Midwest	46	22	32
West	43	24	33
South	32	37	31

Source: Daily Kos / Research 2000.

The results in the South show a wide divergence from those in the rest of the nation. The number of people saying "No" to this question was far higher than in other areas—more than twice what it was in the Northeast. Why should there be such a difference? Table 14.2 presents the results to another question from the same Research 2000 poll.

TABLE 14.2. Research 2000 Poll
on Barack Obama, Regional Breakdown

Do you believe that Barack Obama was born in the United States?

REGION	YES	NO	NOT SURE
Northeast	94	4	3
Midwest	90	6	4
West	87	7	6
South	47	23	30

Source: Daily Kos / Research 2000.

In the results from this question, the southern states are even more sharply divided from the rest of the nation. In fact, following the pattern used by Gallup, you could title these results "only half of Americans living in the South believe that Barack Obama was born in the United States." Though surely many factors are at work, the one most likely responsible for the difference between these states and the rest of the nation is also the one most obvious: race. The question on continental drift was written to draw an emotional response from those uneasy about racial issues—and it found that response.

The way the question is framed is reason enough to disregard the results of this poll as a means of measuring American awareness of continental drift, but there's an even bigger problem. Not only was the question presented in a way that was designed to draw an emotional response, it also pushes the idea that scientific theory is a matter of belief. This framework promotes the impression that data are worth more when they're more popular. This kind of wording not only garners bad responses; it encourages bad interpretation of the results. If 90 percent of Americans believe in the photoelectric effect, but only 20 percent believe in quantum theory, what effect does that have on the performance of your iPhone or the accuracy of your GPS? Absolutely none.

Asking the question in terms of personal belief encourages a complete misinterpretation not only of this particular field of science, but of science in general. What you get when a question is asked this way is a blur of many different results: Are you aware of continental drift? Is continental drift an accurate explanation? Do you like what continental drift has to say about the way forces shape our planet? In short, a belief question is totally worthless, *worse* than worthless, in measuring Americans' knowledge on the subject or the value of the theory. It's foremost a measure of a theory's popularity. And science is not a popularity contest. To treat it as such is to promote the idea that a popular theory is one that is worthy of endorsement, and funding, and research. And vice versa: an unpopular theory should be discounted. Allowing belief to judge fact encourages sloppy thinking, sometimes with serious consequences.

Finally, the presentation of the results in the headline is particularly egregious. "Only 4 in 10 Americans . . ." may reflect the 42 percent who voted "yes" in the poll, but it completely ignores the fact that this is the most popular answer in the survey, and by a broad margin. A much better description of the results would be "More Americans believe that Africa and America were merged in the past than do not believe it." This result could accurately be described as a solid plurality.

Try this alternate headline:

A PLURALITY OF AMERICANS SAY
AFRICA AND AMERICA WERE ONCE UNITED

This is, in all ways, a more accurate presentation of the numbers reflected in the poll. The issues in how the question was phrased are not solved, but at least it addresses the problems with how the data were presented. The original "only 4 in 10" wording is a gross misuse of the raw numbers and places the values in a form designed to make Americans appear more ignorant and to make the theory of continental drift look more controversial. The headline, as originally written, is designed to get the results broadly reported, with no concern about whether they are *correctly* reported.

The results show that only a quarter of Americans don't believe this well-supported theory, while a third of the population admits that they don't understand well enough to have an opinion—a result that should shock no one. It is, in fact, the kind of number you should expect when asking about any scientific theory. How many people walking along your street have a definitive opinion on the mortality of the proton, or even on the Second Law of Thermodynamics?

So why do this? Why purposely ask a poorly framed question about science? Why purposely present the results in a way that makes Americans appear ignorant and ill-informed?

That's a question that should go to Gallup.

Table 14.3 provides a closer look at the poll that generated those headlines for Gallup in the dry season between elections.

TABLE 14.3. Gallup Poll on Evolution

Do you, personally, believe in the theory of evolution, do you not believe in the theory of evolution, or don't you have an opinion either way?

YES	NO	NO OPINION
39%	25%	36%

On the eve of the 200th anniversary of Charles Darwin's birth, a new Gallup Poll shows that only 39% of Americans say they "believe in the theory of evolution," while a quarter say they do not believe in the theory, and another 36% don't have an opinion either way.

Source: Gallup.

Notice that Gallup reinforces the "only-ness" of that 39 percent in the first sentence of the text provided with the poll. However, just as with the poll on continental drift, belief in evolution was the *most popular response*, with "no opinion" a close second. The other response, the response that says "I don't believe in evolution"? That was dead last. In fact, over 50 percent more people said they did believe in evolution than said they didn't. Is that the result you would expect from a headline screaming "only 4 in 10 believe in evolution?"

The results of the poll are best expressed in the same way as the results on the continental drift question: a plurality of Americans say they do believe in evolution. Americans who say they believe in evolution far outnumber those who say that they don't. Not quite what "only 4 in 10" would suggest. Even that statement is of extremely dubious value, considering the nature of the question, but at least it's somewhat accurate in representing the results. Alternatively, Gallup could have gone with this headline:

ONLY 1 IN 4 AMERICANS DOES NOT BELIEVE IN EVOLUTION

This representation is just as accurate as the headline they did run, every bit as "true." However, this version gives a far different view of America's relationship to this issue. If it all seems little more than semantics, well, it is semantics. But semantics have a tremendous impact. Had Gallup gone with one of these alternative headlines—particularly with the "only 1 in 4 Americans does not believe in evolution" choice—do you think film studios would look at a biography of Charles Darwin as being quite so "controversial"? Gallup manufactured controversy in the way it collected and presented its data, turning the plurality of Americans into an isolated minority.

Presented in this way, the data confirm the story we've been told for years. Americans don't believe in evolution. That kind of secular humanist science folderol might be all well and good for those atheists in the rest of the world, by gum, but here in the

United States, we hold fast to Biblical beliefs and just don't cotton to that Darwinist hokum.

So . . . why? Why does Gallup (among others) insist on asking meaningless questions, and then present those results in a sensationalized way that frames Americans as ignorant and anti-science? Why put forward these results on evolution as something extraordinary, when asking about any scientific theory can (as was easily demonstrated by asking a single other science question) generate similar numbers?

Possibly Gallup does this simply because it gets headlines. After all, being in the news is Gallup's business, and we don't hold elections every day. Or maybe that's giving them too much credit. Maybe the Gallup organization itself is too lacking in basic scientific knowledge to see the numerous flaws in what they're asking and how they're presenting it. It's hard to be sure.

Maybe we should have a poll.

Two hundred years after the birth of Darwin, far more Americans accept the fact of evolution of organisms than don't—even if the media are fond of presenting the data quite differently. Among scientists, the acceptance of natural selection as the driving force behind evolution is all but universal. The steady accumulation of knowledge over a century and a half has shown Darwin's ideas to be insightful and accurate. Darwin himself is rightfully celebrated as a figure absolutely central to modern science.

And yet *Darwinism* persists as a label that provides instant condemnation of almost any activity. Even progressives, who understand and acknowledge that organisms evolve in accord with Darwin's theory, still use the term as a pejorative. For the right, *Darwinism* is used as both a label for and explanation of a godless, merciless, death-centric view of both nature and man.

On the left, *Darwinism* can be shorthand for cruelty and the worst aspects of dog-eat-dog commercialism.

It's not hard to see why the most rigid religious followers— those who continue to oppose the teaching of evolution in any form—would attempt to tie Darwin to every sin imaginable. But the reason for the general disdain is more complicated and reflects all the history of how Darwin's ideas have been used and misused. Presented in October of 2009 with the fantastic news of the discovery of *Ardipithecus*, multiple news outlets referred to "the missing link," a nonsensical idea left over from the medieval Great Chain of Being that never appears in Darwin's work. As economic failures loomed in 2008, many economic writers invoked Darwin's name while turning to ideas that had their origin in the writings of Herbert Spencer.

The public understanding of evolution is today still shaped as much by Haeckel's racism and Galton's eugenics as it is by natural selection. Darwinism is associated with the worst aspects of laissez-faire economics: justifying abuse of the downtrodden, programs of forced sterilization, and even the horrors of the Holocaust.

This is a tragedy.

Over a century ago, the Social Darwinists appropriated Darwin's name, but they left most of his theory behind. Their goal then was remarkably like that of the twenty-first-century opponents of evolution: promote ideas intended to cement positions of class and race while eliminating the competition. This pervasive, distorted view of evolution isn't just wrong, it's dangerous.

This view is dangerous because systems facing selective pressure are not restricted to biology acting over millions of years. Selective pressures and the changes they engender are common, nearly universal, aspects of life. The bitter memory of the Social Darwinists and the general misunderstanding of what Darwin discovered do more than just keep us embroiled in pointless battle—they rob us of some of our most effective tools and telling metaphors for describing the world around us.

Misunderstanding Darwin, selective pressure, and evolution cripples us as a society in ways we can't afford. Whether the subject is as serious as national defense, health care, and the economy, or as seemingly trivial as the brands of consumer goods we buy, selective pressures have a lot to say about how the world around us operates—and serious predictions about the consequences of the actions we take.

LIST OF ILLUSTRATIONS
AND THEIR SOURCES

PAGE 51   [Figure 4.1] Darwin's initial sketch of branching
            evolution, 1837. *Source: The Complete Work of Charles
            Darwin Online,* http://www.darwin-online.org.uk/.

PAGE 66   *Left:* Alfred Russel Wallace in 1908, Linnean Society.
            Right: Charles Darwin in 1881, photograph by Her-
            bert Rose Barraud. *Sources:* (left) Linnean Society;
            (right) Milner, "Charles Darwin," *Scientific American*
            (November) 1995: 78.

PAGE 78   Memorial to Those Who Passed Through the Work-
            houses, West Midlands, England. *Source:* Photograph
            by Tony Hisgett, 2009. Licensed under Creative
            Commons and used by permission of the artist.

PAGE 90   The Pedigree of Man. *Source:* Ernst Haeckel, *Gene-
            relle Morphologie der Organismen* (1874).

PAGE 103   [Figure 7.1] *Cirripedia. Source:* Ernst Haeckel,
            Kunstformen der Natur (Artforms of Nature),
            1904, plate 57: Cirripedia. *Source:* Wikimedia
            Commons, http://commons.wikimedia.org/wiki/
            Kunstformen_der_Natur.

PAGE 108   Announcement of the Third International Eugenics
            Congress, 1932. Robert M. Yerkes Papers, Manu-
            scripts & Archives, Yale University, http://media4
            .its.yale.edu/students/sam/MSSA/intellectuals/
            baum/01baum.html.

PAGE 124   Jaffa Gate, Jerusalem. *Source:* "Jerusalem (El-Kouds),
            Jaffa Gate." G. Eric and Edith Matson Photograph
            Collection, 1908. From the United States Library of
            Congress's Prints and Photographs Division under
            the ID matpc.06543.

PAGE 136   Steller's Sea Cow. *Source:* Rev. Henry Neville
            Hutchinson, *Extinct Monsters and Creatures of Other
            Days: A Popular Account of Some of the Larger Forms
            of Ancient Animal Life* (London: Chapman and Hall,
            1883); Second Edition 1896, Plate XXVI, page 248.
            Reprinted by Kessinger Publishing, 2005.

PAGE 148   Wilderness Road. Source: Wikimedia Commons,
Library of Congress, 1755, http://en.wikipedia.org/
wiki/File:Wilderness_road.jpg.

PAGE 160   American Wildcat. *Source:* John James Audubon, *The*
Viviparous Quadrupeds of North America, 1845–1848,
80. New York Public Library Digital Gallery.

PAGE 164   [Figure 12.1] Size Comparison of Largest Sauropods.
Source: Matt Martyniuk, 2007, http://en.wikipedia
.org/wiki/User:Dinoguy2.

PAGE 172   Evolution of the Horse. *Source:* Alex Brollo, 2006,
Wikimedia Commons, http://commons.wikimedia
.org/wiki/File:Horseevolution.png.

PAGE 176   [Figure 13.1] *Archaeopteryx lithographica. Source:*
Wikimedia Commons, http://commons.wikimedia
.org/wiki/File:Naturkundemuseum_Berlin_-_
Archaeopteryx_-_Eichst%C3%A4tt_edit2.jpg.
© Raimond Spekking / Wikimedia Commons /
CC-BY-SA-3.0 & GDFL.

PAGE 181   [Figure 13.2] Chart of Horse Evolution. *Source:*
Author.

PAGE 188   Charles Darwin. Photograph by Elliot & Fry, 1881.
Wikimedia Commons, http://commons.wikimedia
.org/wiki/File:Darwin_1881.jpg.

NOTES

CHAPTER 1: TIME AND MONEY

1. James Hutton, *An Investigation of the Principles of Knowledge and of the Progress of Reason, from Sense to Science and Philosophy,* Vol. 2 (1794; Facsimile edition. London: Thoemmes Press, 1999), 665.

2. James Hutton, *Theory of the Earth,* Vol. 1 (1788), 304.

CHAPTER 2: AND ALL'S RIGHT WITH THE WORLD

1. Paraphrased from Jean-Baptiste Lamarck, *Philosophie Zoologique* (1809), 119.

CHAPTER 3: UNIFORMLY CATASTROPHIC

1. Charles Darwin, *On the Origin of Species,* 6th Edition (1872), Chapter 7, 258. Note: this text is not in the 1866 edition.

2. Ibid.

CHAPTER 4: THIS CURIOUS SUBJECT

1. Charles Darwin, *The Voyage of the Beagle,* 1st Edition. Note: This text is not found in later versions. According to the Museum of Learning,

> In the first edition regarding the similarity of Galápagos wildlife to that on the South American continent, Darwin remarks "The circumstance would be explained, according to the views of some authors, by saying that the creative power had acted according to the same law over a wide area" in a reference to Charles Lyell's ideas of "centres of creation." He notes the gradations in size of the beaks of species of finches, suspects that species "are confined to different islands," "But there is not space in this work, to enter into this curious subject."
> Later editions hint at his new ideas on evolution.

See Museum of Learning, "The Voyage of the Beagle: Publication of FitzRoy S Narrative and Darwin S Book," http://www.museum stuff.com/learn/topics/The_Voyage_of_the_Beagle::sub::Publica tion_Of_FitzRoy_S_Narrative_And_Darwin_S_Book.

2. *The Darwin Correspondence Project,* Letter 441 from Emma Wedgwood to Charles Darwin, November 1838, http://www.darwin project.ac.uk/entry-441.

3. Ibid.

4. Robert Chambers, *Vestiges of the Natural History of Creation,* (1844), Chapter 12, 150. See also the *Unofficial Stephen Jay Gould Archive,* which has the complete text of this and other 19th century works at www.stephenjaygould.org.

5. Robert Chambers, *Vestiges of the Natural History of Creation* (1844), Chapter 12, 152.

6. Charles Darwin's memorial of Anne Elizabeth Darwin, 1851. The original manuscript is in the Darwin Archive of Cambridge University Library (DAR 210.13); a copy of the text is available at The Darwin Project, http://www.thedarwinproject.com/.

7. Francis Darwin, *The Life and Letters of Charles Darwin* (New York: Appleton, 1919), 466.

8. James L. Reveal, F.L.S., Paul J. Bottino and Charles F. Delwiche, "The Darwin-Wallace 1858 Evolution Paper," prepared

for the Department of Cell Biology and Molecular Genetics, University of Maryland. http://www.plantsystematics.org/reveal/PBIO/darwin/darwin02.html.

CHAPTER 5: SOMETIMES THEY EVEN TALK ALIKE

1. Rev. Thomas R. Malthus, *Essay on the Principle of Population* (London: Ward, Lock & Co., 1890 edition), 7.

2. Leonard Huxley, *The Life and Letters of Thomas Henry Huxley,* Vol. 1 (London: Macmillan, 1900), 189.

CHAPTER 7: FLIM-FLAM MEN

1. Charles Darwin, *The Descent of Man*, 2nd Edition (London: John Murray & Co., 1890), 182.

2. Ernst Haeckel, *The History of Creation* (New York: D. Appleton & Co., 1914), Vol. 2, 429.

3. Ernst Haeckel, *The Evolution of Man*, Translated from the 5th edition (Whitefish, MT: Kessinger Publishing, 2004), 1.

4. Martin Gardner, *Fads and Fallacies in the Name of Science* (Mineola NY: Dover Publications, 1957), 229.

CHAPTER 8: BIG MIKE AND THE PAPER HANGER

1. *The Correspondence of Charles Darwin: 1851-1855*, edited by Fredrick Burkhardt and Sydney Smith (Cambridge: Cambridge University Press, 1989), 149–150.

2. "Extracts from F. Galton's Presidential Address to the Anthropological Institute," *Nature 33* (Nov 1885 to Apr 1886) (London: Macmillan and Co, 1886), 297.

3. Correspondence between Charles Darwin and Francis Galton, Letter 412 to Francis Galton. http://galton.org/letters/darwin/correspondence.htm.

4. Francis Galton, "Heredity Intelligence," *The Eclectic Magazine of Foreign Literature, Science, and Art* 59 (1873): 299.

CHAPTER 9: CAMEL PUREE

1. Matthew 19:23-24.

2. I can hear some of you shouting at the page, "Uh uh, birds are dinosaurs and birds survived, so . . . " But while it might be intellectually fun to scrub the word *birds* off the front of *Peterson's Field Guide* and keep that life list of dinosaur sightings, most of us make a distinction between the small feathered creatures of today and the large terrestrial beasts of yore.

CHAPTER 11: IF YOU GO OUT IN THE WOODS TODAY

1. Ronald Reagan, speech to the Western Wood Products Association while he was a candidate for governor of California, *San Francisco Chronicle*, March 12, 1966.

2. Lou Cannon, *Governor Reagan: His Rise to Power* (New York: PublicAffairs, 2003), 178.

3. Henry C. Carey, *The Harmony of Interests* (Philadelphia: Henry Carey Baird, 1868), 228.

CHAPTER 14: GALLUPING SCIENCE

1. Gallup, "On Darwin's Birthday, Only 4 in 10 Believe in Evolution," February 11, 2009. http://www.gallup.com/poll/114544/darwin-birthday-believe-evolution.aspx.

2. Texas State Board of Education, Texas Essential Knowledge and Skills for Science, May 8, 2009, Chapter 112, Subchapter C.

3. Mark Sumner, "Polling Science," *Daily Kos* (August 1, 2009). http://www.dailykos.com/storyonly/2009/7/31/20439/0486.

REFERENCES

Those works listed without a publisher are generally available from multiple sources.

Bakker, Robert. *The Dinosaur Heresies*. New York: Zebra Books, 1986.

Bannister, Robert. *Social Darwinism: Science and Myth in Anglo-American Social Thought*. Philadelphia: Temple University Press, 1979.

Baxter, Stephen. *Ages in Chaos*. Weidenfled & Nicholson, Great Britain, 2003.

Berry, Andrew, ed. *Infinite Tropics: An Alfred Russel Wallace Anthology*. London and New York: Verso, 2002.

Burkhardt, Fredrick and Sydney Smith, eds. *The Correspondence of Charles Darwin: 1851-1855*. Cambridge: Cambridge University Press, 1989.

Cannon, Lou. *Governor Reagan: His Rise to Power*. New York: PublicAffairs, 2003.

Carey, Henry C. *The Harmony of Interests*. Philadelphia: Henry Carey Baird, 1868.

Chambers, Robert. *Vestiges of the Natural History of Creation*, 1844.

Cuvier, Georges. *Discourse on the Revolutionary Upheavals on the Surface of the Earth*, 1812.

Darwin, Charles. *The Voyage of the Beagle*, 1839.

Darwin, Charles. *On the Origin of Species*, 1859.

Darwin, Charles. *The Descent of Man, and Selection in Relation to Sex*, 1871.

Darwin, Charles. "What Mr. Darwin Saw in his Voyage Round the World in the Ship 'Beagle'." New York: Harper, 1898, c1879. University of Missouri Library, Special Collections.

Darwin, Francis. *The Life and Letters of Charles Darwin*. New York: Appleton, 1919 (originally published 1887).

Darwin, Francis, ed. *The Autobiography of Charles Darwin*, 1887.

Desmond, Adrian. *The Politics of Evolution*. Chicago: University of Chicago Press, 1989.

Freinkel, Susan. *American Chestnut: The Life, Death, and Rebirth of a Perfect Tree*. Berkeley and Los Angeles: University of California Press, 2007.

Galton, Francis. *Hereditary Genius*, 1869.

Gardner, Martin. *Fads and Fallacies in the Name of Science*. Mineola NY: Dover Publications, 1957.

Gould, Stephen Jay. *The Structure of Evolutionary Theory*. Cambridge, MA and London: Belknap Press, 2002.

Haeckel, Ernst. *The History of Creation*, 1868.

Haeckel, Ernst. *The Evolution of Man*, 1874.

Huff, Darrell. *How to Lie with Statistics*. New York: Norton, 1954.

Hutton, James. *Theory of the Earth*, 1788.

Hutton, James. *An Investigation of the Principles of Knowledge and of the Progress of Reason, from Sense to Science and Philosophy*, Vol. 2. 1794. Facsimile edition. London: Thoemmes Press, 1999, Vol. 2, 665.

Huxley, Leonard. *The Life and Letters of Thomas Henry Huxley*. Vol. 1. London: Macmillan, 1900.

Johnson, Steven. *Everything Bad Is Good for You*. New York: Riverhead Books, 2005.

Johnson, Steven. *The Ghost Map*. New York: Penguin Group, 2006.

Kemp, T. S. *The Origin and Evolution of Mammals*. Oxford and
    New York: Oxford University Press, 2005.

Lamarck, Jean-Baptiste. *Philosophie Zoologique*. Paris: Librairie F.
    Savy, 1809; Translated by Hugh Elliot. Rosamond, CA: Bill Huth
    Publishing, 2006.

Larson, Edward J. *Evolution: The Remarkable History of a Scientific
    Theory*. New York: Modern Library, 2004.

Lyell, Charles. *Principles of Geology*, Vols. 1–3, 1830–1833.

Malthus, Rev. Thomas R. *Essay on the Principle of Population*.
    London: Ward, Lock & Co., 1890 edition; originally printed for
    J. Johnson in St. Paul's Church-Yard, 1798.

Marsh, Othniel Charles. "Birds with Teeth." Report for Department
    of Interior–U.S. Geological Survey, 1883. Reprint, Whitefish,
    MT: Kessinger Publishing, 2009.

Martin, Paul S. *Twilight of the Mammoths*. Berkeley and Los Ange-
    les: University of California Press, 2005.

McCarren, Mark J. *The Scientific Contributions of Othniel Charles
    Marsh*. Yale University Peabody Museum of Natural History,
    Special Publication No. 15, 1993.

McFadden, Bruce J. *Fossil Horses*. Cambridge: Cambridge Univer-
    sity Press, 1992; paperback, 2008.

McGowan, Christopher. *The Dragon Seekers: How an Extraordi-
    nary Circle of Fossilists Discovered the Dinosaurs and Paved the
    Way for Darwin*. Cambridge, MA: Perseus Publishing, 2001.

Nichols, Peter. *A Voyage for Madmen*. New York: Harper Collins,
    2001.

Packard, Alpheus S. *Lamarck, the Founder of Evolution: His Life
    and Work*. New York: Longmans, Green & Co., 1901; Reprint,
    Middlesex, England: Wildhern Press, 2008.

Pollan, Michael. *The Botany of Desire*. New York: Random House,
    2001.

Repcheck, Jack. *The Man Who Found Time: James Hutton and
    the Discovery of the Earth's Antiquity*. Cambridge, MA: Perseus
    Books, 2003.

Ruse, Michael and Robert J. Richards, eds. *The Cambridge Com-
    panion to the "Origin of Species."* New York: Cambridge Univer-
    sity Press, 2009.

Rushkoff, Douglas. *Life Inc.: How the World Became a Corporation and How to Take It Back*. New York: Random House, 2009.

Simpson, Jacqueline and Steve Roud. *A Dictionary of English Folklore*. Oxford and New York: Oxford University Press, 2000.

Smith, Adam. *The Wealth of Nations*, 1776.

Spencer, Herbert. *Social Statics*, 1851; Revised edition, Honolulu, Hawaii: University Press of the Pacific, 2003.

Sumner, Mark. "Polling Science." *Daily Kos* (August 1, 2009). http://www.dailykos.com/storyonly/2009/7/31/20439/0486.

Valades, Fray Diego de. *Rhetorica Christiana*. Perugia, Italy, 1579.

Wallace, Alfred Russel. *My Life,* 1905.

Wallace, David Rains. *Neptune's Ark*. Berkeley and Los Angeles: University of California Press, 2007.

Wells, H. G. *The Time Machine*, 1895.

Young, James Harvey. *The Medical Messiahs: A Social History of Health Quackery in Twentieth-Century America*. Princeton, NJ: Princeton University Press, 1992.

INDEX

ABOUT THE AUTHOR

Mark Sumner is the award-winning author of several novels, including *Devil's Tower*, nominated for both the Nebula and the World Fantasy Award. His *News from the Edge* series of science fiction mystery novels was the basis for the 2001 television series *The Chronicle*. He worked for several years as a field geologist discovering miles of caves beneath western Kentucky, uncovering dinosaurs in South Dakota, and exploring for minerals across Montana and Wyoming. He now works in an office that is not nearly so scenic or exciting, but does tend to stay warm and dry.

He is a contributing editor at *Daily Kos*, where he frequently writes on issues of science, energy, and the environment. He holds a master's degree from Washington University and lives with his wife in a drafty log cabin near St. Louis, Missouri. He still owns a somewhat rickety and aging ultralight plane for those occasions when the world seems a little too safe.